T. Nakashizuka (Ed.)

Sustainability and Diversity of Forest Ecosystems

An Interdisciplinary Approach

Reprinted from *Ecological Research* Vol. 22 (3) 2007

T. Nakashizuka (Ed.)

Sustainability and Diversity of Forest Ecosystems

An Interdisciplinary Approach

Reprinted from *Ecological Research* Vol. 22 (3) 2007

With 36 Figures

 Springer

Tohru Nakashizuka, Ph.D.
Graduate School of Life Sciences, Tohoku University, 6-3 Aoba, Aramaki, Aoba-ku,
Sendai 980-8578, Japan

Library of Congress Control Number: 2007929726

ISBN 978-4-431-73237-2 Springer Tokyo Berlin Heidelberg New York

Printed on acid-free paper

Springer is a part of Springer+Business Media
springer.com
© The Ecological Society of Japan
Printed in Japan

Typesetting: Scientific Publishing Services Ltd., India
Printing and binding: Hicom, Japan

Ecol Res (2007) 22: 359–360
DOI 10.1007/s11284-007-0356-1

Tohru Nakashizuka

An interdisciplinary approach to sustainability and biodiversity of forest ecosystems: an introduction

Published online: 12 April 2007
© The Ecological Society of Japan 2007

Biodiversity and sustainability of forests

Loss of biological diversity has been noted as an important global issue since the Convention of Biological Diversity in 1992. Biodiversity is decreasing at the fastest rate in the history of the earth, mostly due to human activities. Biodiversity is closely related to the soundness of an ecosystem, and humans benefit from various ecosystem services (Millennium Ecosystem Assessment 2005). Thus, the sustainable use of ecosystems allowing maintenance of biological diversity is an urgent problem that must be solved.

Among terrestrial ecosystems, forests support the richest biological diversity. The interaction of humans and forests has a long history, although recent changes have been the most drastic ever. The rapid decrease and deterioration of forest ecosystems has been caused by social, economic, and ecological factors, which vary locally but are sometimes globally common. However, the drivers and mechanisms causing biodiversity loss through forest utilization and the influences of biodiversity loss are still virtually unknown, although such knowledge is crucial to developing sustainable management strategies (Szaro and Johnston 1996).

Some traditional forest utilization systems are thought to be sustainable, maintaining ecosystem services. However, such traditional systems are in danger of collapsing due to the recent rapid changes in human lifestyles before scientific evaluations of their sustainability are possible (Berkes and Folke 1998). Evaluation of such traditional systems is urgently needed to determine new methods of forest use. Also, the economic considerations of ecosystems and biodiversity have become increasingly important, and their sustainable management can now be analyzed in various ways (Heal 2000).

Thus, the study of sustainable use of forests and biodiversity requires investigation of both human impacts on biodiversity and the impacts of biodiversity change on ecosystem services. The problems are very complex, both ecologically and socially, and thus predictability is usually low. We must elucidate the sources of unpredictability to achieve sustainability of the systems (Levin 1999). Studies should be interdisciplinary and extend beyond the traditional field of ecology.

The symposium and the RIHN project

Given the situation on sustainable forest use outlined above, we started the RIHN (Research Institute for Humanity and Nature) project "Sustainability and Biodiversity Assessment of Forest Utilization Options" in 2002. We have attempted to evaluate the sustainability of forest utilization from various viewpoints, with particular emphasis on biodiversity. The driving forces and incentives causing recent changes in forest utilization systems were investigated, as well as the ecosystem services that may be lost with decreasing biodiversity. Socio- and environmental economic aspects were also assessed for all forest utilization systems, including traditional and so-called sustainable systems in the region.

The papers featured here were presented at an international symposium entitled "Sustainability and Biodiversity of Forest Ecosystems: an Interdisciplinary Approach," held in Kyoto, Japan, on 18 October 2005. Some of the results obtained by the RIHN project are presented, together with other international activities and trends in related issues. Ecologists, forestry scientists, environmental economists, and sociologists joined together to discuss the issue.

T. Nakashizuka
Research Institute for Humanity and Nature,
Kamikamo, Kyoto 603-8047, Japan

Present address: T. Nakashizuka
Graduate School of Life Sciences,
Tohoku University, Aoba 6-3,
Aramaki, Aoba-ku, Sendai 980-8578, Japan
E-mail: toron@mail.tains.tohoku.ac.jp
Tel.: +86-22-7956696
Fax: +86-22-7956699

The contents of this feature

This special feature consists of two parts: (1) forest utilization and its impacts on biodiversity and (2) the basis and practice of sustainable management of forests and biodiversity. The impacts of forest use on biological diversity are reported for plants (Hermy and Verheyen 2007) and insects (Makino et al. 2007). Some of these impacts, in turn, affect human activities such as agriculture (Agetsuma 2007). The influence of forest fragmentation on genetic diversity is also considered (Isagi et al. 2007).

As a practical way to attain sustainable resource use, quality labeling may be effective and is currently undergoing trial utilization (Lagan et al. 2007). Some necessities for the sustainable use of biological resources are discussed (Tsur and Zemel 2007; Akao and Farzin 2007). Also, some traditional systems of local peoples include more sustainable management of resources than current systems (Ichikawa 2007). These reports represent only a small fraction of the studies needed to establish a sustainable system of ecosystem utilization, although they have many implications for further studies. The field of sustainable use needs to be enlarged to include ecology, and the Ecological Society of Japan should be an active participant.

Acknowledgements I thank Dr. Yoh Iwasa, editor of Ecological Research, for agreeing to publish this special feature. RIHN supported several studies published in this feature and the symposium in which all papers published here were presented.

References

Agetsuma N (2007) Ecological function losses caused by monotonous land use induce crop raiding by wildlife on the island of Yakushima, southern Japan. Ecol Res 22(3). doi:10.1007/s11284-007-0358-z

Akao K, Farzin YH (2007) When is it optimal to exhaust a resource in a finite time? Ecol Res 22(3). doi:10.1007/s11284-007-0363-2

Berkes F, Folke C (1998) Linking social and ecological systems. Management practices and social mechanisms for building resilience. Cambridge University Press, Cambridge, pp 459

Heal G (2000) Nature and the market place. Island Press, Washington, pp 203

Hermy M, Verheyen K (2007) Legacies of the past in the present-day forest biodiversity: a review of past land-use effects on forest plant species composition and diversity. Ecol Res 22(3). doi:10.1007/s11284-007-0354-3

Ichikawa M (2007) Degradation and loss of forest land and land-use changes in Sarawak, East Malaysia: a study of native land use by the Iban. Ecol Res 22(3). doi:10.1007/s11284-007-0365-0

Isagi Y, Tateno R, Matsuki Y, Hirao A, Watanabe S, Shibata M (2007) Genetic and reproductive consequences of forest fragmentation for populations of *Magnolia obovata*. Ecol Res 22(3). doi:10.1007/s11284-007-0360-5

Levin S (1999) Fragile dominion. Perseus, Cambridge, pp 250

Makino S, Goto H, Hasegawa M, Okabe K, Tanaka H, Inoue T, Okochi I (2007) Degradation of longicorn beetle (Coleoptera, Cerambycidae, Disteniidae) fauna caused by conversion from broad-leaved to man-made conifer stands of *Cryptomeria japonica* (Taxodiaceae) in central Japan. Ecol Res 22(3). doi:10.1007/s11284-007-0359-y

Millennium Ecosystem Assessment (2005) Ecosystems and human well-being: biodiversity synthesis. World Resources Institute, Washington DC, 86 pp

Lagan P, Mannan S, Matsubayashi H (2007) Sustainable use of tropical forest by reduced-impact logging in Deramakot Forest Reserve, Sabah, Malaysia. Ecol Res 22(3). doi:10.1007/s11284-007-0362-3

Szaro RC, Johnston DW (1996) Biodiversity in managed landscapes: theory and practice. Oxford University Press, New York, pp 778

Tsur Y, Zemel A (2007) Bio-economic resource management under threats of environmental catastrophes. Ecol Res 22(3). doi:10.1007/s11284-007-0361-4

CONTENTS

Part 1
Forest utilization and its impacts on biodiversity

Ecol Res (2007) 22: 361–371
DOI 10.1007/s11284-007-0354-3

Martin Hermy · Kris Verheyen

Legacies of the past in the present-day forest biodiversity: a review of past land-use effects on forest plant species composition and diversity

Received: 4 February 2006 / Accepted: 15 July 2006 / Published online: 20 March 2007
© The Ecological Society of Japan 2007

Abstract Particularly in the temperate climate zone many forests have, at some moment in their history, been used as agriculture land. Forest cover is therefore often not as stable as it might look. How forest plant communities recovered after agriculture was abandoned allows us to explore some universal questions on how dispersal and environment limit plant species abundance and distribution. All studies looking at the effects of historical land use rely on adequate land use reconstruction. A variety of tools from maps, archival studies, and interviews to field evidence and soil analyses contribute to that. They allow us to distinguish ancient from recent forests and many studies found pronounced differences in forest plant species composition between them. A considerable percentage of our forest flora is associated with ancient forests. These ancient forest plant species (AFS) all have a low colonization capacity, suggesting that dispersal in space (distance related) and time (seed bank related) limit their distribution and abundance. However recent forests generally are suitable for the recruitment of AFS. There is clear evidence that dispersal limitation is more important than recruitment limitation in the distribution of AFS. Dispersal in time, through persistent seed banks, does not play a significant role. Ancient forests are not necessary more species-rich than recent forest, but if diversity is limited to typical forest plant species then ancient forests do have the highest number of plant species, making them highly important for nature conservation. The use of molecular markers, integrated approaches and modelling are all part of the way forward in this field of historical ecology.

Keywords Historical ecology · Dispersal limitation · Recruitment limitation · Conservation · Forest cover · Research needs · Future research

M. Hermy
Division Forest, Nature and Landscape,
Department of Land Management and Economics,
Catholic University of Leuven, Celestijnenlaan 200 E,
3001 Leuven, Belgium
E-mail: martin.hermy@biw.kuleuven.be

K. Verheyen (✉)
Laboratory of Forestry, Ghent University,
Geraardsbergse Steenweg 267, 9090 Melle-Gontrode, Belgium
E-mail: kris.verheyen@ugent.be

Introduction

Global change, referring both to climatic and land-use change, has attracted enormous interest in both science and policy and its impact worries politicians, ecologists and conservationists. An increasing number of scientific papers is devoted to the consequences of climatic change. However, at the same time we tend to forget that these changes interfere with huge land-use changes. Between 1990 and 2000, the Food and Agriculture Organization (FAO) of the United Nations (2001) estimated the yearly worldwide deforestation at 14.6 million ha, 97.2% of which occurred in the tropical zone. In non-tropical areas, losses were largely compensated by afforestation both spontaneously (2.6 million ha year^{-1}) and through planting (0.7 million ha year^{-1}). Afforestation mainly occured on agricultural land. Given the extent of afforestation, research on the recovery of forest plant communities is an important topic both for managers and ecologists (Flinn and Vellend 2005). The intended and natural restoration of forest communities may also yield insights into fundamental questions on how plant species disperse and recruit in these new forests. But as forest species colonization is generally slow, it will take many decades or centuries before the results of the recent afforestations will be visible.

However, afforestation is not new. Today many landscapes worldwide continue to bear the imprint of historical land-use changes. Particularly across northwestern Europe and northeastern North America phases of forest clearance were followed by agricultural

Fig. 1 Changes over time in forest cover in northwestern Belgium (former county of Flanders, calculated from Tack et al. 1993) **a** and eastern US, **b** (all states east of the Mississippi River, from Flinn and Vellend 2005). Forest cover estimates are approximate between 7000 BC and 400 AD in Flanders and before 1700 in US

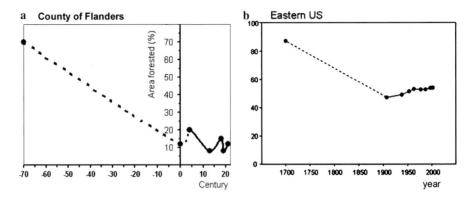

abandonment and forest regeneration (Fig. 1). While in northeastern North America the first large-scale deforestation largely took place during the 18th and particularly the 19th century (e.g., Foster 1992; Hall et al. 2002; Flinn and Vellend 2005, but see Burgess and Sharpe 1981), in much of northwestern Europe deforestation goes back much further in time (e.g., Rackham 1980; Tack et al. 1993; Verheyen et al. 1999). In some regions, for example the county of Flanders in Belgium, several phases of abandonment and reforestation occurred (Fig. 1), resulting in a complex pattern of forest patches of different age. Furthermore, in some regions forests that have developed on farmland since the 19th century may represent as much as 80% of the current forest cover (Foster 1992; Foster et al. 1998; Grashof-Bokdam and Geertsema 1998). In Flanders only about 23,000 ha (16%) of the present-day forests already existed in 1770 (De Keersmaeker et al. 2001). As expected, these recent forests do differ in vegetation and soils from forests that were never cleared (so called primary forests, sensu Peterken 1974, 1981) or from forests that already existed before a certain threshold date (so-called ancient forests, sensu Hermy et al. 1999). Such observations lead ecologists to ask whether recent forests will ever recover from the past deforestation and the impact from former agricultural land use (Dupouey et al. 2002; Flinn and Vellend 2005), and if so, how long will it take? And if not, do we need to intervene, for instance by actively introducing forest plant species?

Here we want to review some of the results of empirical studies done since the 1950s in both northeastern North America and northwestern Europe. Furthermore we will embark on the basic ecological questions raised by the latter on dispersal and recruitment limitation. We will thereby focus on herbaceous understory plants of deciduous forests (cf. Flinn and Vellend 2005), as in many cases the woody layer in these forests has been strongly manipulated and trees have often planted by men, limiting the conclusions that can be drawn with respect to the ongoing demographic processes for woody species. We finish with some suggestions on new directions for research.

Land use reconstruction

All studies looking at the effects of historical land-use change rely on adequate reconstructions of history. Historical land-use maps, aerial photographs in combination with written sources and archaeological research may document the change in forest cover over time. They will show large differences between northwestern Europe and northeastern North America, for example (Fig. 1). In the latter large-scale deforestation mainly started early in the 19th century and forest cover progressively decreased to a minimum in about the end of the 19th or early 20th century, after which it increased to the current forest cover of somewhere between 20 and 50% (Smith et al. 2004; Flinn and Vellend 2005) or even up to 80% (Foster 1992; Hall et al. 2002). In Europe deforestation often goes back much further in time. In the northwestern part of Belgium (i.e., the former county of Flanders) for example, a considerable forest area was already lost from the Neolithicum onwards (Tack et al. 1993) and reached a first minimum during the Roman era (about 12% of the land had forest cover at that time). Afterwards it significantly increased during the dark ages, probably reaching about 20% around 400 AD. Then a new deforestation phase took place, reaching a new low point around 1,300 when all peat reserves in that region were exhausted. As a result, the value of fuel wood increased tremendously and agricultural land was afforested again. A new maximum was reached around the end of the 18th century. Then coal gradually took over as an energy source and forest cover declined again. However, gradually more wood was needed in coal mining and forest area increased and this was later combined with renewed multifunctional interest in forests to reach its actual cover of about 12% in this region. Although this shows the overall long-term trend in forest cover, for detailed more spatially explicit information about forest-cover changes, data are mostly only reliable from the 18th century onwards (but see Verheyen et al. 1999).

The criterion for distinguishing primary from secondary forests is forest continuity. Primary forest has never been reclaimed, but the status is often difficult to prove. So instead ancient and recent forest were used as terms (Peterken 1981). Ancient forest is then defined as forest that has existed continuously since at least a specified date (threshold date), selected on the availability of historical land-use information and differing between studies and countries (see Fig. 2, Hermy et al. 1999). So using historical sources one can distinguish remnants of ancient forest, mostly managed, from recent forests, also managed but restored on former farmland. It also enables one to estimate the age of recent forest patches (Peterken 1981; Smith et al. 1993; Hermy et al. 1999). The use of threshold dates to distinguish ancient from recent forest stands is particularly essential in northwestern Europe with its complex land-use history (cf. Fig. 1). It may mean that at least some part of the ancient forest is in fact secondary in origin (Fig. 2). So to be sure about the actual history of a parcel of forest land-use maps should be complemented with interviews with landowners, observations of field evidence such as stone and earth walls, tree fall pits and mounds, open-grown trees (Rackham 1980; Peterken 1981; Marks and Gardescu 2001), and with historical information from

other approaches such as palynology and archeology (e.g., Dupouey et al. 2002; Vanwalleghem et al. 2004). Chemical soil analysis, such as the measurement of the total phosphate content, may be a helpful tool indicating former agricultural land use, even if it dates back from Roman times or earlier (see, e.g., Provan 1971; Gebhardt 1982; Wulf 1994; Wilson et al. 1997; Honnay et al. 1999; Verheyen et al. 1999). However, even then uncertainties about the exact nature and scale of land-use changes remain (Whitney 1994; Verheyen et al. 1999; Hall et al. 2002). The increased and obvious use of geographical information systems (GIS) in historical land-use reconstruction, creating accurate and detailed maps, may therefore create an illusion of certainty.

Differences in flora, plant traits and colonization capacity

After abandonment, former agricultural land is gradually colonized by herbaceous plant species favouring open habitats. A rapid succession of ruderal to highly competitive plant species (sensu Grime 1979) is observed and within 5–20 years pioneer tree and shrub species (e.g. *Pinus* spp., *Betula* spp., and *Salix* spp.) may start to dominate and shade out open-habitat species. In both the northwestern European and northeastern North American temperate zone within a time span of about 100 years the pioneer tree species are gradually replaced by later successional tree genera such as beech (*Fagus*), oak (*Quercus*), lime (*Tilia*) or maple (*Acer*) (e.g. Ellenberg 1978; Packham and Harding 1982). However, long after the reconstitution of the tree layer in recent forest, considerable differences with ancient forests in the composition and abundance of particularly herbaceous species persist (e.g. Rackham 1980; Hermy and Stieperaere 1981; Peterken and Game 1984; Whitney and Foster 1988; Matlack 1994; Petersen 1994; Wulf 1997; Bellemare et al. 2002, for a review see Hermy et al. 1999; Verheyen et al. 2003a). From a review of the European literature Hermy et al. (1999) found that about 30% of all forest plant species are confined to ancient forests [so-called ancient forest plant species (AFS)]. AFS seem to exhibit a common ecological profile (Table 1): they tend to be more shade-tolerant than other forest plant species, avoid dry and wet sites and tend to be more stress-tolerant

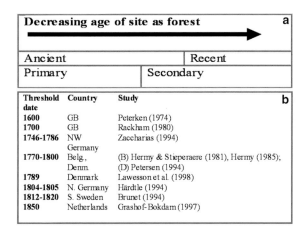

Fig. 2 The relationships between primary, secondary, and ancient and recent forests in Europe (**a**) (cf. Peterken 1981). Ancient forests is defined as forest that has existed continuously since at least a specified date (threshold date), selected on the availability of historical land-use information and differing between studies and countries (**b**) (cf. Hermy et al. 1999)

Table 1 Ancient forest plant species share a number of common characteristics (summarized from Hermy et al. 1999; Verheyen et al. 2003a)

Hermy et al. (1999): AFS	Verheyen et al. (2003a) added: AFS
Preferentially occur on rich, moist but not wet soils	Have relatively large seeds
Prefer weakly acid to neutral soils	Do not form no persistent seed banks
Have intermediate N availability	Require specific germination requirementsconditions
Are preponderance of geophytes and hemicryptophytes	Have delayed age of first reproduction
Mainly are summergreen spp.	Predominantly show vegetative reproduction (clonal growth)
24% are considered: myrmecochores	Have limited fecundity
Mainly show a stress-tolerant plant strategy	

than other forest plant species that belong more to the competitive plant strategy type. But, it also became clear that considerable regional variation in AFS existed, indicating that in some regions a species may be typical of ancient forests, while in others it may be capable of colonizing recent forests. This suggests that AFS are not a black-white or all-or-nothing phenomenon; they may all be able to colonize to some extent. Therefore another, more species-specific approach is needed. Based on an extensive quantitative study of all available high quality data sources from both northeastern North America (eight sources) and northwestern Europe (12 sources) Verheyen et al. (2003a) suggested calculating per species a so-called colonizing capacity index (CCI):

$$\mathrm{CCI}_i = \left[\frac{(1.5\mathrm{RE}* + \mathrm{RE})(1.5\mathrm{AN}* + \mathrm{AN})}{\mathrm{AN}* + \mathrm{AN} + \mathrm{RE}* + \mathrm{RE}} \right] \times 100/1.5.$$

RE* means the number of studies in which species i is statistically significantly more frequent in recent forest; RE refers to the number of studies in which species i is equally or more frequent in recent forest; AN* represents the number of studies in which species i is statistically significantly more frequent in ancient forest and AN is then the number of studies in which species i is more frequent in ancient forest.

The index ranges from $+100$ (species strongly associated with recent forest) to -100 (species strongly associated with ancient forest) and thus expresses the colonization capacity of an individual species: the lower the value, the lower the colonization capacity. Table 2 gives an overview of the forest plant species with the lowest colonization capacity. Apart from relatively rare taxa, such as *Carex pendula*, *Ranunculus lanuginosus*, *Melica uniflora*, *Corydalis cava* in NW Europe, it also includes relatively common species such as *Lamiastrum galeobdolon*, *Carex sylvatica*, *Circaea lutetiana* in NW Europe. Although not formally analyzed, some closely related species (e.g. *Allium ursinum* vs. *A. tricoccum*, *Asarum europaeum* vs. *A. canadense*, *Paris quadrifolia* vs. *Trillium* spp., *Polygonatum multiflorum* vs. *Polygonatum* spp.) exhibit similar responses on both continents. Yet, this work also found some remarkable, yet unrevealed continental contrasts within the same species, for example, *Oxalis acetosella* (CCI Europe -75, 87 in N. America) and *Circaea lutetiana* (CCI Europe -67 vs. N. America 0). More importantly, plant trait correlation structure was similar in the European and North American datasets. Species having relatively large seeds, low fecundity, unassisted (short-distance) dispersal, specific germination requirements, delayed age of first reproduction, clonal growth and no persistent seed bank generally proved to be slow colonizers (see Table 1 and also Whigham 2004). Many of these life-history traits may make AFS more sensitive to habitat loss and fragmentation (Flinn and Vellend 2005) and therefore also prone to extinction (cf. Peterken 1977).

In the examples given above ancient forest has been defined as forest existing since at least the 19th century (see Fig. 2). But what if the deforestation occurred much earlier, and thus the threshold date goes back much further, e.g. to Roman times? Perhaps unexpectedly, in the northeast of France Dupouey et al. (2002) showed that even 2,000 years after reforestation forests still exhibit clear floristic differences compared to forests that were never cleared. Similar, but less convincing, results

Table 2 Forest plant species most strongly associated with ancient deciduous forests of northwestern Europe and northeastern North America

European herb species (174 species; $n = 12$)	CCI	American herb species (44 species; $n = 8$)	CCI
Carex pallescens	-100	*Cardamine diphylla*	-100
Carex pendula	-100	*Chimaphila maculata*	-100
Hypericum pulchrum	-100	*Clintonia borealis*	-100
Luzula sylvatica	-100	*Coptis trifolia*	-100
Lysimachia nemorum	-100	*Caulophyllum thalictroides*	-87
Lysimachia vulgaris	-100	*Claytonia caroliniana*	-87
Scilla non-scripta	-100	*Osmorhiza claytoni*	-87
Succisa pratensis	-100	*Trillium erectum*	-87
Veronica montana	-100	*Trillium undulatum*	-87
Galium odoratum	-95	*Viola macloskeyi*	-87
Corydalis cava	-83	*Asarum canadense*	-80
Lathyrus sylvestris	-83	*Uvularia perfoliata*	-80
Ranunculus lanuginosus	-83	*Actaea rubra*	-67
Lamiastrum galeobdolon	-79	*Allium tricoccum*	-67
Anemone nemorosa	-77	*Hepatica acutiloba*	-67
Oxalis acetosella	-75	*Medeola virginiana*	-67
Paris quadrifolia	-75	*Viola rotundifolia*	-67
Carex sylvatica	-74	*Geranium maculatum*	-33
Melica uniflora	-71	*Aster acuminatus*	-20
Viola reichenbachiana	-71	*Cypripedium acaule*	-20
Circaea lutetiana	-67	*Erythronium americanum*	-20
Euphorbia dulcis	-67	*Galium circaezans*	-20
Primula elatior	-67	*Podophyllum peltatum*	-20
Luzula pilosa	-59	*Polygonatum biflorum*	-20

Values show the colonizing capacity indices (CCI), which range from -100 (strongly associated with ancient forest) to $+100$ (strongly associated with recent forests) (recalculated from Verheyen et al. 2003a) n number of studies were found by Vanwalleghem et al. (2004) in central Belgium.

Why do forest plants fail to colonize new sites? Recruitment versus dispersal limitation

The strong association of many forest plant species with ancient forest suggests that dispersal in space (distance related) and time (seed bank related) may be limiting the distribution and abundance of many forest plant species. But it might also be that the habitat of recent forest is unsuitable for the recruitment of these species, as the latter has been considerably modified by the former agricultural land use. Particularly in Europe, a long history of management also adds to these effects (Fig. 3). Basically all forests, whether ancient or recent, have been managed as coppice, coppice-with-standards or high forest.

Does land-use affect environmental conditions?

Agriculture potentially affects vegetation both directly, by locally eliminating plants and their diaspores, and indirectly, by altering environmental conditions. Numerous researchers have indeed found that soils under recent forest have higher pH, nutrient concentrations and lower organic matter than soils under ancient forests (e.g. Koerner et al. 1997; Bossuyt et al. 1999a; Honnay et al. 1999; Verheyen et al. 1999; Flinn et al. 2005). However the magnitude and persistence of these effects shows variation between regions (see Verheyen et al. 1999; Koerner et al. 1997 vs. Bellemare et al. 2002; Compton and Boone 2000; Flinn et al. 2005) and between elements (compare mobile basic cations vs. persistent phosphate; e.g. Honnay et al. 1999). If differences persist to some extent in recent forests, and for total phosphate differences may persist for even 2,000 years (cf. Dupouey et al. 2002), this may suggest that recruitment is limiting the presence of AFS.

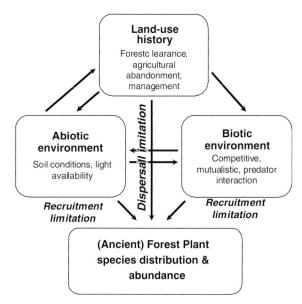

Fig. 3 Diagram showing the direct and indirect effects of land-use history on forest plant species distribution and abundance (adapted from Flinn and Vellend 2005)

Recruitment limitation

Tests of this hypothesis may come from two sources (Flinn and Vellend 2005): research investigating the performance of forest plant species and in particular AFS in recent forests—as many of the AFS do show some colonization—and studies monitoring experimental introductions of forest plant species in recent sites. Contrary to expectations, performance in recent forests has often been similar or even been better (Table 3, cf. Endels et al. 2004). Similar results were found in central Massachusetts by Donohue et al. (2000) for the

Table 3 Performance of two forest plant species in an alluvial recent and ancient forest (Langerode, central Belgium) (adapted from Endels et al. 2003)

	Ancient forest			Recent forest		
	n	mean $\pm$ SE	*	n	mean $\pm$ SE	*
Primula elatior						
No. of flowers per plant	179	15.79 $\pm$ 1.18	a	86	43.9 $\pm$ 7.03	b
No. of fruits per plant	110	9.75 $\pm$ 0.87	a	39	16.9 $\pm$ 4.24	b
Total no. of seeds per plant	110	338.19 $\pm$ 38.25	a	39	516.2 $\pm$ 155.76	b
Germination %	110	33.75 $\pm$ 3.54	a	40	35.6 $\pm$ 5.89	a
Geum urbanum						
No. of flowers per plant	93	5.18 $\pm$ 0.31	a	140	7.5 $\pm$ 0.37	b
No. of fruits per plant	93	4.01 $\pm$ 0.27	a	140	5.5 $\pm$ 0.30	b
Total no. of seeds per plant	93	394.18 $\pm$ 32.30	a	137	626.4 $\pm$ 39.17	b
Germination %	91	69.88 $\pm$ 2.75	a	135	71.1 $\pm$ 2.50	a

Primula elatior is usually considered an ancient forest plant species and *Geum urbanum* is not. The significance of differences using the Mann–Whitney U test is indicated by different letters. Recent forest: planted after 1945; ancient forest: forest since at least 1775
* Values are averages, and differences between two forest types that are significantly different are indicated with different letters

evergreen, woody clonal understory species, *Gaultheria procumbens*. In contrast, Vellend (2005) found that for *Trillium grandiflorum* plants of the same age were smaller and less likely to flower in recent forests than in ancient forests in central New York. Yet, only very few studies are available for a very limited number of species. So more work is needed here.

In experiments where forest plant species have been introduced, seed sowing and adult transplants in recent forests usually proved successful (e.g. Petersen and Philipp 2001; Heinken 2004; Verheyen and Hermy 2004; Graae et al. 2004). These all seem to suggest that recruitment may not be the prime problem (cf. Ehlrén and Eriksson 2000). However no long-term studies on the performance of the introduced plants in recent forests are available. As many forest plant species have long life spans (e.g. Inghe and Tamm 1985; Cain and Damman 1997)—Ehrlén and Lethilä (2002) inferred an average life span of 64 year for forest field layer species as opposed to 22 year for species from open habitats—really long-term studies are needed. At the same time, it seems to suggest that dispersal limitation may be the main problem.

Dispersal limitation in space and in time

Dispersal in plant species covers two aspects: dispersal in space, referring to dissemination of diaspores away from sexually reproductive parent plants and often categorized as dispersal modes based on the morphology of dispersal units, and dispersal in time, referring to the ability of plant species to build seed banks in the soils. These again may take two forms: transient versus persistent seed banks. The former persist in the soil for less than one year and the latter persist in the soil for at leastone year (Bakker 1989; Thompson et al. 1997).

Dispersal in space, or the transport of diaspores away from a parent, is an essential prerequisite for recruitment and as such is an important step in the colonization process (see also Fig. 7). It is to be expected that dispersal limitation will increase with the increasing isolation of recent from ancient forest. In the available literature two useful extremes in isolation may be encountered: (1) recent forests stands immediately bordering ancient forest stands on the same soils and with similar light conditions and (2) recent forests isolated from ancient forests by a hostile—mostly agricultural—landscape matrix. In the former, the available studies came up with colonization rates between ± 20 and 100 m century^{-1} depending on the plant species (northeastern North America: Matlack 1994; southern Sweden: Brunet and von Oheimb 1998; central Belgium: Bossuyt et al. 1999b; Honnay et al. 1999). These rates are all based on the furthest individual and in all studies it is assumed that colonization started at the edge of the ancient forest. Colonization rates are thus generally low and difficult to reconcile with the high rates of migration found after the last glacial period (cf. Cain et al. 1998).

Colonization success in isolated recent forest patches, defined as the ratio of occupied target fragments (in the example given, between 30 and 42 years old) over the sum of the number of occupied target fragments and the number of suitable but not occupied target fragments, proved to be very low for 85% of the forest plant species and strongly decreased with increasing distances. However, colonization success was clearly linked to the connectivity of the landscape (Fig. 4; Honnay et al.

Fig. 4 Probability of occurrence of forest plant species in a forest patch as a function of the average distance to the five nearest source populations of that species in two landscapes differing in connectivity and forest cover. *Triangles*, for the more-fragmented landscape and filled circles for the high-connectivity landscape. **a** *Arum maculatum*, **b** *Geum urbanum*, **c** *Adoxa moschatellina*, **d** *Primula elatior* (from Honnay et al. 2002a)

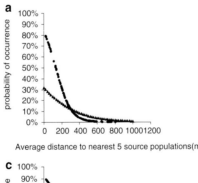

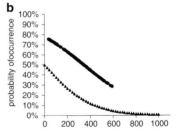

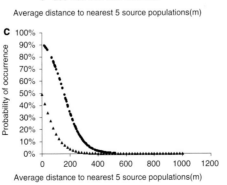

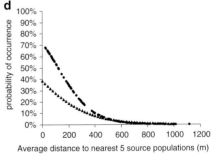

2002a). In high-connectivity landscapes (in the example, 42% forest cover) colonization success was clearly higher than in low-connectivity landscapes (with 7% forest cover).

Although colonization rates of most forest plant species were on average less than a few metres per year, one must be careful about generalizations. Indeed, chance events may disperse plant diaspores over long distances. It all relates to the importance of the tail of the seed shadow (cf. Cain et al. 1998; Higgins and Richardson 1999; Cain et al. 2000). Indeed, Cain and colleagues could explain the high Holocene colonization rates of the North American forest herb *Asarum canadense* based on the tail of the seed dispersal curve of the species.

Given the more or less unique response of individual forest plant species, one might expect a relation between species dispersal mode and colonization success. Endo- and epizoochores are generally clearly more successful in colonizing recent forests than the other groups (cf. Honnay et al. 2002a; Dzwonko and Loster 1992; Matlack 1994; Grashof-Bokdam and Geertsema 1998; Brunet and von Oheimb 1998; Bellemare et al. 2002; Takahaskhi and Kamitani 2004). Dispersal distances of seeds transported by ants in Japanese deciduous forest have been reported to range from 0.28 to 0.64 m (Ohara and Higashi 1987; Higashi et al. 1989; Ohkawara and Higashi 1994; Ohkawara et al. 1996). However, others (e.g. Mabry et al. 2000; Singleton et al. 2001; Ito et al. 2004; Wulf 2004) found no association between dispersal mode and colonization ability. Although an intuitively attractive and logical approach, attempts to relate dispersal mode to colonization ability have not always been successful, mainly because dispersal classes are based on seed morphology and may poorly represent realized dispersal modes and distances (Vellend et al. 2003). Some myrmeco- and barochores do occur frequently in secondary stands (e.g. *Trillium erectum*, *Carex* spp. in W. Massachusetts, Bellemare et al. 2002, *Viola riviniana*, *Glechoma hederacea* in W. Europe) and conversely some anemochores (e.g. *Adiantum pedatum* in W. Massachusetts, Bellemare et al. 2002, *Pteridium aquilinum*, *Dryopteris carthusiana* in W. Europe) are strongly confined to ancient forests. This suggests that additional empirical data on dispersal of forest plant species are needed, and the potential of perhaps unusual dispersal events, such as vertebrate (including men) dispersal of baro- and myrmecochores should not be overlooked (cf. Bellemare et al. 2002).

Deforestation and transformation to agricultural land use completely destroys the forest plant communities. Yet if diaspores of forest plant species form persistent seed banks, they could as least survive a temporary agricultural land use if it lasts less than a few decades. So dispersal in time could overcome a temporal agricultural land-use phase. However, many forest plant species and most of the AFS do not form a persistent seed bank (cf. Brown and Oosterhuis 1981; Bierzychudek 1982; Bossuyt and Hermy 2001; Bossuyt

et al. 2002; Verheyen et al. 2003a, Table 1). By definition their transient seed banks do not allow the recovery of ancient forest plant communities even if agricultural land use is only for a few years. So, seed banks are mostly of no significance in the restoration of forest plant communities, at least for the AFS (Bossuyt et al. 2002; Honnay et al. 2002b). In contrast species of edges and clear cuts often do have persistent seed banks (e.g. Ash and Barkham 1976; Brown and Oosterhuis 1981; Bossuyt and Hermy 2001; Bossuyt et al. 2002). The number of seed bank papers specifically related to the impact of past land use of forests is however scarce, and hence we still have little insight on how past land-use changes remain visible in the seed banks of forests.

Are ancient forests more diverse than recent forests?

As AFS are dispersal limted one could expect that recent forest stands are less species-rich than ancient forests. Peterken (1974) showed that ancient forest species are quality indicators and their diversity is a means to estimate the nature conservation value of forests. Yet as there is a lot of regional variation in AFS it is difficult to compare across regions and comparisons therefore depend on how species pools or on how the sampling units (plots vs. forest patches) are defined. Our group (e.g. Tack et al. 1993; Jacquemyn et al. 2001; Butaye et al. 2001) used a list of 203 plant species that are restricted to forests, including species typical of forest interiors, forest edges and woody species that are usually not planted. Others used the entire flora of forest patches (e.g. Peterken and Game 1984); still others used plots (e.g. Hermy 1985; Motzkin et al. 1996; Bellemare et al. 2002; Flinn and Marks 2004). If total plant species richness is used, ancient forests plots are not necessary richer in plant species than recent forests (cf. Hermy 1985). If some quality aspect is introduced (e.g. AFS, or typical forest plant species), then ancient forests patches/plots usually are more species rich than recent forests (Fig. 5), at least if the environmental conditions are comparable (cf. Graae 2000). This makes ancient forests important hot spots for forest quality indicating plant species and an important issue in conservation. We essentially need to conserve forests for forest species and not for open habitat species. However, forests may be important for open-habitat species as well, particularly if associated with ancient open habitats such as rides (Peterken and Francis 1999). Ancient forests have been assigned as past-natural forests (Peterken 1977), although often having a simple, artificial structure due to management, but having a composition reflecting those of natural forests. So the conservation value of ancient forests, even with their artificial structure, is highly important, as their recreation within a few hundreds of years is unfeasible. Conversion to conifer or other exotic species stands must be avoided as it is usually a way of no return (cf. Peterken 1981).

368

Are ancient forest plant species affected by traditional management practices?

Many of the ancient forests in Europe have been converted from coppice and coppice-with-standards to some form of high forest. The fundamental feature of the coppice system is the restocking method: each coppice stand grows mainly from shoots that spring from the cut stumps of the previous stand; rotation usually occurs between 5 and 30 years, depending on the regeneration rate of species and the demand for wood products. Coppices mixed with trees (so called standard trees) grown on a multiple of the underwood rotation (usually about 100 years) are indicated as coppice-with-standards. Coppice-with-standards therefore traditionally yielded three main products, namely sticks and brushwood from the underwood, timber from standard trees scattered amongst the underwood, and pasture from the herbaceous field layer and the grassy rides and clearings (Peterken 1981). Ancient forests if combined with

ancient coppice stools and trees—as is particularly often the case in northwestern Europe (Rackham 1980; Peterken 1981; Tack et al. 1993)—are an important cultural heritage and reflect the values society put on these ecosystems in the past. As many ancient forests have been managed for centuries under these traditional management schemes, it may have influenced their composition and therefore also the distribution and abundance of many of the AFS. The regular cutting of the coppice—typically with a rotation cycle of 5–30 years—causes a temporary increased light intensity from about 5% of full daylight to virtually 100% (cf. Buckley 1994). Over the years, this—together with the cutting itself—has favoured those species that are tolerant to this treatment. This may have enhanced the abundance of a number of particularly spring flowering plant species in northwestern Europe such as. *Primula elatior, Viola riviniana, Cardamine pratensis* and *Anemone nemorosa* (cf. Rackham 1975; Ash and Barkham 1976; Peterken 1981) (Fig. 6). It is clear that these effects are included in most

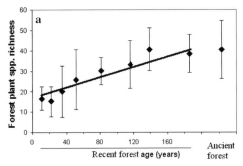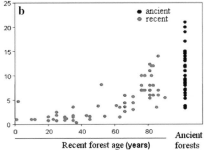

Fig. 5 Forest plant species richness increases with forest age and ancient forests do have on average more typical forest plant species. **a** Forest patch data from central Belgium; average values per age class, *error bars* and fitted regression line (adapted from:

Jacquemyn et al. 2001); **b** forest plot (herb species richness per 180 m²) data from central New York (figure from: Flinn and Vellend 2005)

Fig. 6 Response between 1972 and 1978 of *Primula elatior*, a typical spring-flowering perennial plant species of ancient forests on rich soils, on coppicing in the winter of 1971–1972 and regrowth on coppice stools is shown as well. *Circles* indicate the number of flowers and *triangles* the number of inflorescences (adapted from Rackham 1975)

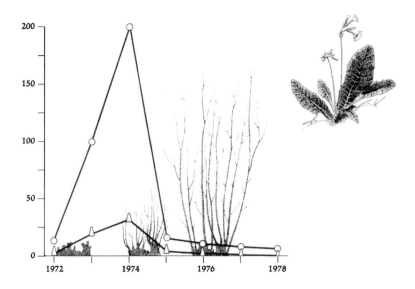

10

comparisons of the flora of ancient and recent forests, particularly in Europe as there is here a long tradition of coppice or coppice-with-standards management. Until now, field experience suggests that management effects are mainly reflected in the abundance of some AFS; carpets of spring vernals are indeed typical of many ancient coppiced forests and far less typical for high forest management systems. Whether management also affected presence–absence patterns of AFS remains unclear and it is not clear what the exact long-term effects will be of the conversion of traditional systems to high forest management.

Recent developments and the way forward

Since the first papers on ancient forest plant species in the 1950s (for a review see Hermy et al. 1999), much progress has been made in elucidating the effects of past land use on today's patterns of (ancient) forest plant species distribution and diversity. The dispersal of diaspores in recent forests appears to be the most critical step limiting the whole colonization process from seed to adult. Recruitment does play a role, but overall it is of secondary influence, and in fact also the second big step in the whole colonization process (Fig. 7). Results to date also point to some important directions for further research (see also Flinn and Vellend 2005): (1) elaborating the impact studies of past land use on forest composition and diversity of

regions other than North America and northwestern Europe such as Japan (see Ito et al. 2004) and NE China, (2) elaborating the approach to other taxa than plants (e.g. Desender et al. 1999), (3) the need for integrated approaches for explaining the distribution and abundance of forest plant species (cf. Verheyen et al. 2003b) and the comparison of results from multiple landscapes with different land-use histories (cf. Vellend 2003; Vellend et al. 2006), (4) the use of molecular markers to document the impacts of land-use history on genetic variation or the origin of colonists (cf. Jacquemyn et al. 2004; Vellend 2004; Honnay et al. 2005), (5) the role of multitrophic interactions (e.g. mutualism, predation) in determining the establishment of forest plant species, (6) modeling the impact of land-use changes on forest plant species distribution and diversity (cf. Matlack and Monde 2004; Verheyen et al. 2004), (7) quantifying possible extinction debts for ancient forests as a consequence of fragmentation (cf. Vellend et al. 2006). These are not only ecologically sound questions, but they are also important for the conservation of forest plant species and their communities.

Acknowledgments The first author would like to thank prof. Tohru Nakashizuka of the Research Institute for Humanity and Nature (RHIN) for the kind invitation to the international symposium of RHIN on sustainability and biodiversity of forest ecosystems, October 18, 2005. Blackwell Publishing and the Ecological Society of America are acknowledged for permission to reproduce several figures or part of them.

References

Ash JE, Barkham JP (1976) Changes and variability in the field layer of a coppiced woodland in Norfolk, England. J Ecol 64:697-712

Bakker JP (1989) Nature management by grazing and cutting. Kluwer, Dordrecht

Bellemare J, Motzkin G, Foster DR (2002) Legacies of the agricultural past in the forested present: an assessment of historical land-use effects on rich mesic forests. J Biogeogr 29:1401–1420

Bierzychudek P (1982) Life histories and demography of shade-tolerant temperate forest herbs: a review. New Phytol 90:757–776

Bossuyt B, Hermy M (2001) Influence of land use history on seed banks in European temperate forest ecosystems: a review. Ecography 24:225–238

Bossuyt B, Deckers J, Hermy M (1999a) A field methodology for assessing man-made disturbance in forest soils developed in loess. Soil Use Manage 15:14–20

Bossuyt B, Hermy M, Deckers J (1999b) Migration of herbaceous plant species across ancient–recent forest ecotones in central Belgium. J Ecol 87:628–638

Bossuyt B, Heyn M, Hermy M (2002) Seed bank and vegetation composition of forest stands of varying age in central Belgium: consequences for regeneration of ancient forest vegetation. Plant Ecol 162:33–48

Brown A, Oosterhuis C (1981) The role of buried seeds in coppice woods. Biol Conser 21:19–38

Brunet J (1994) Der Einfluss von Waldnutzung und Waldgeschichte auf die Vegetation südschwedischer Laubwälder. Norddeutsche Naturschutzakademie-Berichte 3/94:96–101

Brunet J, Von Oheimb G (1998) Migration of vascular plants to secondary woodlands in southern Sweden. J Ecol 86:429–438

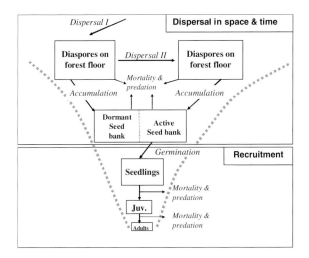

Fig. 7 A conceptual diagram showing the bottle-neck model of the colonization process with a dispersal and a recruitment step. The primary dispersal from the parent plant (dispersal I) may be followed by a secondary dispersal phase (dispersal II). After dispersal, diaspores are incorporated in a seed bank, which may be dormant or active. From that diaspores may germinate and grow up. At each life stage losses may occur through mortality and predation, ultimately yielding a low number of adults (box size relative to number of recruits). The decreasing number of individuals is emphasized by the bottle-neck pattern of *dots*

Buckley GP (ed) (1994) Ecology and management of coppice woodlands. Chapman & Hall, London

Burgess RL, Sharpe DM (eds) (1981) Forest island dynamics in man-dominated landscapes. Springer, Heidelberg

Butaye J, Jacquemyn H, Hermy M (2001) Differential colonization causing non-random forest plant community structure in a fragmented agricultural landscape. Ecography 24:369–380

Cain ML, Damman H (1997) Clonal growth and ramet performance in the woodland herb, *Asarum canadense*. J Ecol 85:883–897

Cain ML, Damman H, Muir A (1998) Seed dispersal and the Holocene migration of woodland herbs. Ecol Monogr 68:325–347

Cain ML, Milligan BG, Strand AE (2000) Long-distance seed dispersal in plant populations. Am J Bot 87:1217–1227

Compton JE, Boone RD (2000) Long-term impacts of agriculture on soil carbon and nitrogren in New England forests. Ecology 81:2314–2330

De Keersmaeker L, Rogiers N, Lauriks R, De Vos B (2001) Ecosysteemvisie bos Vlaanderen, ruimtelijke uitwerking op basis van de natuurlijke bostypes op basis van bodemgroeperingseenheden en historische boskaarten. Report VLINA C97/06, AMINAL, afd. Natuur.109p

Desender K, Ervynck A, Tack G (1999) Beetle diversity and historical ecology of woodlands in Flanders. Belg J Zool 129:139–156

Donohue K, Foster DR, Motzkin G (2000) Effects of the past and the present on species distribution: land-use history and demography of wintergreen. J Ecol 88:303–316

Dupouey JL, Dambrine E, Laffite JD, Moares C (2002) Irreversible impact of past land use on forest soils and biodiversity. Ecology 83:2978–2984

Dzwonko Z, Loster S (1992) Species richness and seed dispersal to secondary woods in southern Poland. J Biogeogr 19:195–204

Ehrlén J, Lethilä K (2002) How perennial are perennial plants? Oikos 98:308–322

Ehrlén J, Eriksson O (2000) Dispersal limitation and patch occupancy in forest herbs. Ecology 81:1667–1674

Ellenberg H (1978) Vegetation Mitteleuropas mit den Alpen, 2nd edn. Ulmer, Stuttgart

Endels P, Adriaens D, Verheyen K, Hermy M (2004) Population structure and adult performance of forest herbs in three contrasting habitats. Ecography 27:225–241

FAO (2001) Global Forest Resources Assessment 2000. Food and Agriculture Organization of the United Nations, Rome.http://www.fao.org/forestry/fo/fra/index.jsp

Flinn KM, Marks PL (2004) Land-use history and forest herb diversity in Tompkins County, New York, USA. In: Honnay O, Verheyen K, Bossuyt B, Hermy M (eds) Forest biodiversity: lessons from history for conservation. CABI, Wallingford, pp 81–95

Flinn KM, Vellend M (2005) Recovery of forest plant communities in post-agricultural landscapes. Front Ecol Environ 3:243–250

Flinn KM, Vellend M, Marks PL (2005) Environmental causes and consequences of forest clearance and agricultural abandonment in central New York. J Biogeogr 32:439–452

Foster DR (1992) Land-use history (1730–1990) and vegetation dynamics in central New England. J Ecol 80:753–772

Foster DR, Motzkin G, Slater B (1998) Land-use history as long-term broad-scale disturbance: regional forest dynamics in central New England. Ecosystems 1:96–119

Gebhardt VH (1982) Phosphatkartierung und bodenkundliche Geländeuntersuchungen zur Eingrenzung historischer Siedlungs- und Wirtschaftsflächen des Geestinsel Flögeln, Kreis Cuxhaven. Probleme Kustenforschung sudlichen Nordseegebiet 14:1–9

Graae BJ (2000) The effect of landscape fragmentation and forest continuity on forest floor species in two regions of Denmark. J Veg Sci 11:881–892

Graae BJ, Hansen T, Sunde PB (2004) The importance of recruitment limitation in forest plant species colonization: a seed sowing experiment. Flora 199:263–270

Grashof-Bokdam C (1997) Forest species in an agricultural landscape in the Netherlands: effects of habitat fragmentation. J Veg Sci 8:21–28

Grashof-Bokdam CJ, Geertsema W (1998) The effect of isolation and history on colonization patterns of plant species in secondary woodland. J Biogeogr 25:837–846

Grime JP (1979) Plant strategies & Vegetation processes. Wiley, Chichester

Hall B, Motzkin G, Foster DR (2002) Three hundred years of forest and land-use change in Massachusetts, USA. J Biogeogr 29:1319–1335

Härdtle W (1994) Zur Veränderung und Schutzfähigkeit historisch alter Wälder in Schleswig-Holstein. Norddeutsche Naturschutzakademie-Ber 3/94:88–96

Heinken T (2004) Migration of an annual myrmecochore: a four year experiment with *Melampyrum pretense* L. Plant Ecol 170:55–72

Hermy M (1985) Ecologie en fytosociologie van oude en jonge bossen in Binnen-Vlaanderen. PhD Thesis, State University Gent

Hermy M, Stieperaere H (1981) An indirect gradient analysis of the ecological relationships between ancient and recent riverine woodlands to the south of Bruges. Vegetation 44:46–49

Hermy M, Honnay O, Firbank L, Grashof-Bokdam C, Lawesson JE (1999) An ecological comparison between ancient and other forest plant species of Europe, and the implications for conservation. Biol Conserv 91:9–22

Higashi S, Tsuyuzaki S, Ohara M, Ito F (1989) Adaptive advantages of ant-dispersed seeds in the myrmecochorous plant *Trillium tschonoskii* (Liliaceae). Oikos 54:389–394

Higgins SI, Richardson DM (1999) Predicting plant migration rates in a changing world: the role of long-distance dispersal. Am Nat 153:464–475

Honnay O, Hermy M, Coppin P (1999) Impact of habitat quality on forest plant species colonization. Forest Ecol Manage 115:157–170

Honnay O, Verheyen K, Butaye J, Jacquemyn H, Bossuyt B, Hermy M (2002a) Possible effects of climate change and habitat fragmentation on the range of forest plant species. Ecol Lett 5:525–530

Honnay O, Bossuyt B, Verheyen K, Butaye J, Jacquemyn H, Hermy M (2002b) Ecological perspectives for the conservation of plant communities in European temperate forests. Biodivers Conserv 11:213–242

Honnay O, Jacquemyn H, Bossuyt B, Hermy M. (2005) Forest fragmentation effects on patch occupancy and population viability of herbaceous plant species. New Phytol 166:723–736

Inghe O, Tamm C-O (1985) Survival and flowering of perennial herbs. IV. The behaviour of *Hepatica nobilis* and *Sanicula europaea* on permanent plots during 1943–81. Oikos 45:400–420

Ito S, Nakayama R, Buckley GP (2004) Effects of previous land-use on plant species diversity in semi-natural and plantation forests in a warm-temperate region in southeastern Kyushu, Japan. Forest Ecol Manag 196:213–225

Jacquemyn H, Butaye J, Hermy M (2001) Forest plant species richness in small, fragmented mixed deciduous forest patches: the role of area, time and dispersal limitation. J Biogeogr 28:801–812

Jacquemyn H, Honnay O, Galbusera P, Roldan-Ruiz I (2004) Genetic structure of the forest herb *Primula elatior* in a changing landscape. Mol Ecol 13:211–219

Koerner W, Dupouey JL, Dambrine E, Benoît M (1997) Influence of past land use on the vegetation and soils of present day forest in the Vosges mountains, France. J Ecol 85:351–358

Lawesson JE, de Blust G, Grashof-Bokdam C, Firbank L, Honnay O, Hermy M., Hobitz P, Jensen LM (1998) Species diversity and area-relationships in Danish beech forests. Forest Ecol Manag 106:235–245

Mabry C, Ackerly D, Gerhardt F (2000) Landscape and species level distribution of morphological and life history traits in a temperate woodland flora. J Veg Sci 11:213–224

Marks PL, Gardescu S (2001) Inferring forest stand history from observational field evidence. In: Egan D, Howell EA (eds) The historical ecology handbook: a restorationist's guide to reference ecosystems. Island Press, Washington

Matlack RG (1994) Plant species migration in a mixed-history forest landscape in eastern North America. Ecology 75:1491–1502

Matlack RG, Monde J (2004) Consequences of low mobility in spatially and temporally heterogeneous ecosystems. J Ecol 92:1025–1035

Motzkin G, Foster D, Allen A, Harrod J, Boone R (1996) Controlling site to evaluate history: vegetation patterns of a New England sand plain. Ecol Monogr 66:345–365

Ohara M, Higashi S (1987) Interference by ground beetles with the dispersal by ants of seeds of *Trillium* species (Liliaceae). J Ecol 75:1091–1098

Ohkawara K, Higashi S (1994) Relative importance of ballistic and ant dispersal in two diplochorous *Viola* species (Violaceae). Oecologia 100:1091–1098

Ohkawara K, Higashi S, Ohara M (1996) Effects of ants, ground beetles and the seed-fall patterns on myrmecochory of *Erythronium japonicum* Decne. (Liliaceae). Oecologia 106:500–506

Packham JR, Harding DJL (1982) Ecology of woodland processes. Arnold, London

Peterken GF (1974) A method for assessing woodland flora for conservation using indicator species. Biol Conserv 6:39–45

Peterken GF (1977) Habitat conservation priorities in British and European woodlands. Biol Conserv 11:223–236

Peterken GF (1981) Woodland conservation and management. Chapman and Hall, London

Peterken GF, Francis M (1999) Open spaces as habitats for vascular ground flora species in the woods of central Lincolnshire, UK. Biol Conserv 91:55–72

Peterken GF, Game M (1984) Historical factors affecting the number and distribution of vascular plant species in the woodlands of central Lincolnshire. J Ecol 72:155–182

Petersen PM (1994) Flora, vegetation, and soil in broadleaved ancient and planted woodland, and scrub on Røsnæs, Denmark. Nordic J Bot 14:693–709

Petersen PM, Philipp M (2001) Implantation of forest plants in a wood on former arable land: a ten year experiment. Flora 196:286–291

Provan DM (1971) Soil phosphate analysis as a tool in archaeology. Norw Arch Rev 4:37–50

Rackham O (1975) Hayley wood. Cambridge and Isle of Ely Naturalists' Trust Ltd., Cambridge

Rackham O (1980) Ancient woodland, its history, vegetation and uses in England. Arnold, London

Singleton R, Gardescu S, Marks PL, Geber MA (2001) Forest herb colonization of post-agricultural forests in central New York, USA. J Ecol 89:325–338

Smith BE, Marks PL, Gardescu S (1993) Two hundred years of forest cover changes in Tompkins County, New York. B Torrey Bot Club 120:229–247

Smith WB, Miles PD, Vissage JS, Pugh SA (2004) Forest resources of the United States, 2002.General technical report NC-241, St.Paul, USDA Forest service, North Central Forest Exp station

Tack G, Van den Bremt P, Hermy M (1993) Bossen van Vlaanderen. Een historische ecologie. Davidfonds, Leuven

Takahaskhi K, Kamitani T (2004) Effect of dispersal capacity on forest plant migration at a landscape scale. J Ecol 92:778–785

Thompson K, Bakker JP, Bekker RM (1997) The soil seed banks of North West Europe: methodology, density and longevity. Cambridge university press, Cambridge

Vanwalleghem T, Verheyen K, Hermy M, Poesen J, Deckers S (2004) Legacies of Roman land-use in the present-day vegetation in Meerdaal forest (Belgium)? Belg J Bot 137:181–187

Vellend M (2003) Habitat loss inhibits recovery of plant diversity as forests regrow. Ecology 84:1158–1164

Vellend M (2004) Parallel effects of land-use history on species diversity and genetic diversity of forest herbs. Ecology 85:3043–3055

Vellend M (2005) Land-use history and plant performance in populations of *Trillium grandiflorum*. Biol Conserv 124:217–224

Vellend M, Myers JA, Gardescu S, Mark PL (2003) Dispersal of Trillium seeds by deer: implications for long-distance migration of forest herbs. Ecology 84:1067–1072

Vellend M, Verheyen K, Jacquemyn H, Kolb A, van Calster H, Peterken GF, Hermy M (2006) Extinction debt of forest plants persists for more than a century following habitat fragmentation. Ecology 87:542–548

Verheyen K, Hermy M (2004) Recruitment and growth of herb-layer species with different colonizing capacities in ancient and recent forest. J Veg Sci 15:125–134

Verheyen K, Bossuyt B, Hermy M, Tack G (1999) The land use history (1278–1990) of a mixed hardwood forest in western Belgium and its relationship with chemical soil characteristics. J Biogeogr 26:1115–1128

Verheyen K, Honnay O, Motzkin G, Hermy M, Foster D (2003a) Response of forest plant species to land-use change: a life-history based approach. J Ecol 91:563–577

Verheyen K, Guntenspergen G, Biesbrouck B, Hermy M (2003b) An integrated analysis of the effects of past land-use on forest plant species colonization at the landscape scale. J Ecol 91:731–742

Verheyen K, Vellend M, Van Calster H, Peterken G, Hermy M (2004) Metapopulation dynamics in changing landscapes: a new spatially realistic model for forest plants. Ecology 85:3302–3312

Whigham D (2004) Ecology of woodland herbs in temperate deciduous forests. Ann Rev Ecol Syst 35:583–621

Whitney GG (1994) From coastal wilderness to fruited plain: a history of environmental change in temperate North America from 1500 to the present. Cambridge University Press, Cambridge

Whitney GG, Foster DR (1988) Overstorey composition and age as determinants of the understorey flora of woods of Central New England. J Ecol 76:867–876

Wilson BR, Moffat AJ, Nortcliff S (1997) The nature of three ancient woodland soils in southern England. J Biogeogr 24:633–636

Wulf M (1994) Überblick zur Bedeutung des Alters von Lebensgemeinschaften, dargestellt am Beispiel "historisch alter Wälder". Norddeutsche Naturschutzakademie-Ber 3/94:3–14

Wulf M (1997) Plant species as indicators of ancient woodland in northwestern Germany. J Veg Sci 8:635–642

Wulf M (2004) Plant species richness of afforestation with different former use and habitat continuity. Forest Ecol Manage 195:191–204

Zacharias D (1994) Bindung von Gefässpflanzen an Wälder alter Waldstandorte im nördlichen Harzvorland Niedersachsens - ein Beispiel für die Bedeutung des Alters von Biotopen für den Pflanzenartenschutz. Norddeutsche Naturschutzakademie-Berichte 3/94:76–88

Ecol Res (2007) 22: 372–381
DOI 10.1007/s11284-007-0359-y

Shun'ichi Makino · Hideaki Goto · Motohiro Hasegawa
Kimiko Okabe · Hiroshi Tanaka · Takenari Inoue
Isamu Okochi

Degradation of longicorn beetle (Coleoptera, Cerambycidae, Disteniidae) fauna caused by conversion from broad-leaved to man-made conifer stands of *Cryptomeria japonica* (Taxodiaceae) in central Japan

Received: 27 January 2006 / Accepted: 26 August 2006 / Published online: 3 April 2007
© The Ecological Society of Japan 2007

Abstract We studied the species richness and assemblages of longicorn beetles (Coleoptera, Cerambycidae,
Disteniidae) in ten secondary broad-leaved stands and
eight plantation stands of Japanese cedar (*Cryptomeria
japonica*) of various ages after clear-cutting or plantation
in Ibaraki, central Japan. The species richness of longicorns, which were collected with Malaise traps, was the
highest in young stands, decreasing with the age of the
stand for both broad-leaved and conifer stands. A
canonical correspondence analysis divided the 18 plots
into three groups based on longicorn assemblages and
environmental variables. These three groups consisted of
(1) very young (1–4 years old) stands after clear-cutting
or plantation; (2) 12- to over 100 year-old broad-leaved
stands; (3) 7- to 76-year-old conifer stands. The species
richness of the longicorns was the highest in the young
stands followed, in order of decreasing species richness, by broad-leaved stands and conifer stands. Possible
causes of the high species richness in young stands include large amounts of coarse wood debris and flowers,
which are resources for oviposition and nutrition for
adults, respectively. The lower longicorn diversity in
conifer stands than in broad-leaved stands may be due
to the lower diversity of trees available as host plants in
the former. Almost all species that occurred in conifer
stands were also collected in young and/or broad-leaved
stands, but the reverse was not true, suggesting that
conifer plantations cannot replace broad-leaved stands
in terms of longicorn biodiversity. We argue that an
extensive conversion of broad-leaved forests into conifer
plantations will lead to an impoverishment of the longicorn fauna, which may result in the degradation of
ecosystem functions possibly carried out by them.

S. Makino (✉) · H. Goto · M. Hasegawa · K. Okabe ·
H. Tanaka · T. Inoue · I. Okochi
Forestry and Forest Products Research Institute,
Matsunosato 1, Tsukuba, Ibaraki 305-8687, Japan
E-mail: makino@ffpri.affrc.go.jp
Tel.: +81-29-8298249
Fax: +81-29-8743720

Keywords Biodiversity · Chronosequence · Forest ·
Insect · Plantation

Introduction

Forestry, a human activity, has affected the natural
habitats of plants and animals in various ways. In particular, plantation forestry often has negative effects on
the biodiversity of various organisms in many parts of
the world because of structural and functional dissimi-
larities between the native and replacement forest (Palik
and Engstrom 1999; Moore and Allen 1999). Warnings
have been given on the need to be careful about the
possible degradation of ecological functions as well as of
biodiversity caused by intensive silviculture associated
with plantation practices (Hartley 2002).

In Japan, one of the largest impacts on habitats by
forestry has been the conversion from natural or secondary broad-leaved forests to plantations of conifers,
mainly of Japanese cedar (*Cryptomeria japonica*) and
hinoki cypress (*Chamaecyparis obtusa*). The area occupied by plantations of these two conifers was greatly
increased after the Second World War to fulfill a high
demand for timber and currently represents about 19%
of the total forest area in Japan (Japan Agriculture
Statistics Association 2001).

Japanese conifer plantations are principally monocultures composed of one of the above two conifer
species. The establishment of monocultures at the ex-
pense of natural mixed-species stands are expected to
reduce the biodiversity of forest-dependent organisms
(e.g. Palik and Engstrom 1999). In fact, the overstory
vegetation of Japanese cedar stands is poorer in biodiversity than natural or secondary forests of a similar age
(Tanaka et al., unpublished data). This may lead to the
impoverishment of biodiversity or, at the very least, to
changes in the assemblage of forest animals that directly
or indirectly depend on plants. To test this hypothesis
and to determine to what extent insect assemblages of

conifer plantations diverge from those of broad-leaved forests, we need to monitor insects along chronosequences of both types of forests because arthropod species richness and assemblages generally change as forests grow (e.g. Buddle et al. 2000; Inoue 2003; Trofymow et al. 2003; Sueyoshi et al. 2003; Makino et al. 2006).

Although a number of studies have been published which compare species richness and assemblage of insects between Japanese cedar or hinoki cypress plantations and broad-leaved forests (e.g. Ohashi et al. 1992; Shibata et al. 1996; Mizota and Imasaka 1997; Maeto and Makihara 1999; Maeto et al. 2002; Maeto and Sato 2004; Sayama et al. 2005), the study plots of these studies consisted of only a few age classes of conifer stands. There is a scarcity of studies that have been conducted with the aim of comparing insect biodiversity between the two types of forests through their chronosequences. To this end, we conducted the present investigation in which we monitored longicorn beetles in both conifer plantations of Japanese cedar and secondary deciduous broad-leaved forests of a wide range of ages.

Longicorn beetles are typical forest-dependent insects because they almost exclusively feed on living, dying or dead trees in the larval stage (Linsley 1959, 1961; Hanks 1999). Saproxylic insects, including most longicorns, thus depend on dead wood and old trees, and they mediate or promote decomposition processes of microorganisms (Grove 2002). Additionally, many longicorn adults visit flowers to feed on nectar and/or pollen (e.g. Linsley 1959, 1961; Kuboki 1987); as such, they act as pollinators for some plant species (Kuboki 1998). These facts suggest that the functions possibly performed by longicorn beetles in forest ecosystems should not be overlooked. If their abundance and/or assemblages are greatly affected by the conversion of a broad-leaved forest to a conifer plantation, a loss or degradation of their functions may follow.

The aim of this study was, therefore, to determine the extent to which the conversion of broad-leaved forests into conifer plantations affects the species richness and assemblages of longicorns in central Japan, and how they respond to chronosequential changes in forest characteristics.

Materials and methods

Study areas

This study was conducted in two areas, Ogawa and Satomi, Ibaraki Prefecture, central Japan. Both areas have plantations of *Cryptomeria japonica* and *Chamaecypress obtusa*, although the percentage of plantation areas is much larger in Satomi than in Ogawa (94 vs. 47% of total forested area; Tanaka et al., unpublished data). Ogawa is located at the southern edge of the Abukuma Mountains in Kitaibaraki (approximately

$36°56'$N, $140°35'$E; 580–800 m a.s.l.). The dominant large trees in deciduous broad-leaved forests in the area are *Quercus serrata*, *Q. mongolica* and *Fagus crenata*. Some areas of the broad-leaved forests in Ogawa have been subjected to human activities, such as burning, cattle grazing and clear-cutting for fuel wood (Suzuki 2002), and the small-scale clear-cutting of broad-leaved stands has been repeatedly carried out up to the present time to collect bed logs for mushroom culture. In addition, following the Second World War, the area converted from broad-leaved stands to conifer plantations greatly increased. These human practices have resulted in a mosaic-like landscape composed of secondary broad-leaved stands and conifer plantations of various ages. We selected ten plots in broad-leaved stands to obtain stands that formed a chronosequence from 1 to over 100 years old after clear-cutting (Table 1). All plots were located within an approximately 30-km^2 area. Satomi (approximately $36°50'$N, $140°34'$E; 700–800 m a.s.l.) is about 10 km southwest of the Ogawa area. We set eight plots of Japanese cedar plantations in an approximately 10-km^2 area in Satomi. The conifer plots also formed a chronosequence from 1 to 76 years old in age after plantation. In Ogawa, the annual mean temperature is 10.7°C and mean annual precipitation is 1910 mm (Mizoguchi et al. 2002).

Forest characteristics

In order to determine how forest characteristics affect longicorn assemblages, we carried out plant censuses in the above plots. We established a line transect at each plot from September 2000 to October 2003. Each transect line was 100 m long and was set to cover the entire variation of the forest. All trees and vines of at least 2 m in height and at least 5 cm in diameter at breast height (DBH) were counted, and their girth at breast height (GBH) was measured in a total of forty 5 × 5-m quadrats along both sides of the 100-m transect line mentioned above; the frequency of trees smaller than 5 cm in DBH in the 40 quadrats was also censused. Forest floor vegetation with a height of less than 2 m (forest floor plants) was censused following the Braun–Blanquet method for a 1 × 1-m subquadrat which was set in each 5 × 5-m quadrat.

Insect collection

We collected longicorn beetles with standard Townes-type Malaise traps (Golden Owl Publishers; 180 cm long, 120 cm wide, 200 cm high) in 2002 in Ogawa and 2003 in Satomi. The traps were placed inside the stand to avoid the edge effect, except in very young, grassland-like plots which had no tall trees. Trapped insects were collected every 2 weeks from late April to early November in both areas. A mixture of ethanol and propylene glycol was used as preservative in the insect

Table 1 Study plots and collection summary of longicorn beetles (Coleoptera, Cerambycidae, Disteniidae)

Site	Plot code	Age (year)	Area (ha)	Number of longicorn species	Number of longicorn individuals	$1-D^{a}$
Ogawa	O1	1	3	52	329	0.946
	O4	4	5	43	339	0.852
	O12	12	4	38	145	0.923
	O24	24	24	39	150	0.941
	O51	51	10	30	144	0.870
	O54	54	14	27	109	0.928
	O71	71	19	33	153	0.935
	O128	>100	98	34	119	0.939
	O174	>100	11	27	106	0.933
	O178	>100	10	35	149	0.930
Satomi	S3	3	4	40	357	0.677
	S7	7	6	24	142	0.840
	S9	9	5	27	55	0.968
	S20	20	5	16	29	0.951
	S29	29	14	18	61	0.922
	S31	31	12	27	93	0.916
	S75	75	3	19	57	0.933
	S76	76	3	12	24	0.935

[a] Simpson's index of diversity (see section analyses of insect assemblages)

containers of the traps. The collected insects were morphologically identified to the species level. Longicorn samples from two traps, 10 m distant from each other, were used for each plot in the following analysis. All voucher specimens are deposited at the Forestry and Forest Products Research Institute.

Analyses of insect assemblages

Data of longicorn beetles were pooled for each plot throughout the season and analyzed for a comparison of species richness or assemblage among the plots. As an index of diversity, we calculated Simpson's index of diversity, which is the complement of D defined as follows (Magurran 2004):

$$D = \sum \frac{n_I(n_I - 1)}{N(N - 1)}$$

where N is the total number of individuals and n_I is the number of individuals of the Ith species. In order to test differences in the number of longicorn species or individuals or in the diversity indices among stands or forest types, the analysis of variance (ANOVA) was carried out with SYSTAT ver. 9.01 for Windows (SPSS 1998) and STATISTICA for Windows (StafSoft Inc. 2000). If data were not normally distributed, they were log-transformed. Canonical correspondence analysis (CCA) was performed with CANOCO for Windows, ver. 4.5 (ter Braak and Smilauer 2002) to make an ordination analysis of longicorn assemblages among the plots and to relate environmental variables (see below) with the assemblages. Only those longicorn species with a total count of at least three individuals were used. As environmental variables for CCA, we used the following plant community indices: species richness of plants for three size classes (trees with DBH ≥ 5 cm; trees with DBH < 5 cm; forest floor plants), the density of trees

with DBH ≥ 5 cm and the maximum and average diameter of trees at breast height. In this analysis, the scores of the first and second axis in a detrended correspondence analysis (DCA) for the plant community group ordination for the three size classes were also used in order to investigate the effects of the composition of the plants species on communities of longicorn beetles. In the DCA of plant communities, species with at least three individuals in total were used, and population data were transformed using logarithmic transformation, $\log_{10}(x + 1)$. In the DCA of trees with DBH ≥ 5 cm, the 1-year-old site (O1) and the 4-year-old site (O4) were excluded because there were no trees of these size classes. Likewise, in the DCA of forest floor plants, the 1-year-old site was excluded. Population data were transformed using logarithmic transformation, $\log_{10}(x + 1)$. Environmental variables were tested by forward selection of variables with the Monte Carlo test using 499 unrestricted permutations ($P < 0.05$).

The indicator species analysis (Dufrêne and Legendre 1997) was performed with PC-ORD ver. 4 (McCune and Mefford 1999) in order to identify the representative species of longicorn beetles for the groups of sites identified during the ordination processes. This analysis produced indicator values (IndVals) for each species in each group of sites, which were subsequently tested for statistical significance using a Monte Carlo technique.

Results

Longicorn species richness

We collected a total of 2561 individuals of 106 longicorn species in two sites (see Appendix): 99 species in ten plots in broad-leaved stands in Ogawa and 66 species in eight conifer plots in Satomi (Table 1). The average number of longicorn species (species richness) was larger in the

375

broad-leaved plots (35.8; SD: 7.7) than in coniferous ones (22.9; SD: 8.7) (ANOVA: $df = 1$, 16; $F = 11.18$; $P = 0.004$). The average number of longicorn individuals was also larger in broad-leaved than in conifer stands (ANOVA, log-transformed data: $df = 1,16$; $F = 7.40$; $P = 0.017$). Simpson's index of diversity ($1-D$), however, was not different between them (ANOVA: $df = 1,16$; $F = 0.02$; $P = 0.891$). In order to compare species richness in forests of different ages, plots of each area were divided into the following three age classes based on years after clear-cutting (for broad-leaved stands) or after plantation establishment (for conifer stands): 1–19 years old, 20–100 years old and over 100 years old (only in Ogawa). In both sites, longicorn species richness differed among the age classes (ANOVA: Ogawa, $df = 2$, 7; $F = 5.03$; $P = 0.044$; Satomi, $df = 1$, 6; $F = 6.03$; $P = 0.05$), with the youngest plots having more species than older ones (Fig. 1). In Ogawa, although pairwise comparisons with Bonferroni adjustment showed that differences were insignificant between any pair of the three age classes ($P > 0.05$), species richness was significantly higher in the youngest class (1–19 years old) than the seven older stands combined (ANOVA: $df = 1$, 8; $F = 11.48$; $P = 0.01$). Simpson's index was not different among the three age classes in Ogawa (ANOVA, $df = 2$, 7; $F = 0.27$; $P = 0.774$) or between the two classes in Satomi (ANOVA: $df = 1$, 6; $F = 2.42$; $P = 0.171$).

Longicorn assemblages

The result of CCA ordination is shown in Fig. 2. Eigenvalues of the first two axes and their cumulative percentage variance of species data were 0.271, 0.218 and 27.5%, respectively. The ordination divided the 18 plots into three groups, which correspond to three dif-

ferent forest types as follows. The first type (initial stage stands; abbreviated as INI) was composed of the two youngest plots in Ogawa (O1, O4), which were open fields of herbs and shrubs after clear-cutting, and of one plot in Satomi (S3), which was a young plantation of seedlings of Japanese cedar. This forest type is thus characterized as plots of "stand initiation stage" (Oliver 1981) after disturbances. The second type (broad-leaved stands; BLD) consisted of eight broad-leaved plots in Ogawa (O12–O178), and the third type (coniferous plantations; CPL) comprised seven conifer plots in Satomi (S7–S76). The number of longicorn species was largest in INI, followed by BLD and CPL (smallest) (Fig. 3). The difference was highly significant among the three forest types (ANOVA: $df = 2$, 15; $F = 24.534$; $P < <0.001$) and between any pair after correction for multiple comparison with the Bonferroni adjustment (Fig. 3). Simpson's index, however, was not significantly different among the three types (ANOVA: $df = 2$, 15; $F = 1.263$; $P = 0.311$), showing that the evenness of longicorn assemblages did not change among them.

In CCA, forward selection for factors related to the species composition of longicorns revealed the following five significant ($P < 0.05$) variables of forest characteristics which are indicated with arrows in Fig. 2: scores of the second axis in DCA of forest floor plants (arrow A); the maximum DBH of trees (B); the number of tree species with DBH $\geq$ 5 cm (C); the number of floor plant species (D); the scores of the first axis in DCA of trees with DBH $\geq$ 5 cm (E). The directions of arrows B and C, both representing the maturation process of forest growth, are similar to that of the first axis of the CCA diagram (Fig. 2) and explain the discrimination in the composition of longicorn species between the two older forest types (BLD and CPL) and the young one (INI). In contrast, arrows A and E, which represent the scores of DCA of forest floor plants and trees, respectively, are more vertical in the CCA diagram; they explain the discrimination in the composition between BLD and CPL. Arrow D, which represents the species richness of the forest floor plants, was short, showing that its effect on the longicorn species composition may be marginal.

Longicorn species which had statistically significant index values (IndVals) with the indicator species analysis were found in the INI and BLD group (Table 2). In contrast, no species had significant IndVals in the CPL group.

Discussion

Species richness and stand age

In various insects, species richness changes with forest age (e.g. Buddle et al. 2000; Inoue 2003; Trofymow et al. 2003; Sueyoshi et al. 2003; Makino et al. 2006). In the present investigation, the species richness and the number of individuals of longicorns were found to show a similar response to the stand age in both the broad-

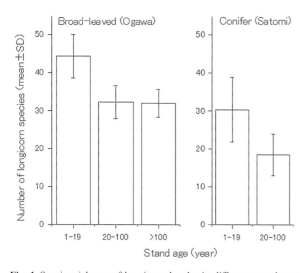

Fig. 1 Species richness of longicorn beetles in different age classes in broad-leaved and conifer plantation stands

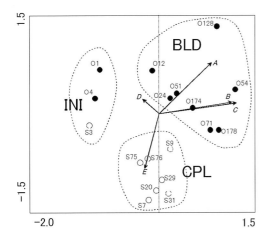

Fig. 2 Ordination of longicorn assemblages by canonical correspondence analysis (CCA) analysis. Three forest types are identified: *INI* initiation stage, *BLD* broad-leaved stands, *CPL* conifer plantations. The *arrows* denoted with *A–E* are vectors of environmental (vegetational) variables which were selected ($P < 0.05$) with forward selection. *A* Scores of the second axis in the detrended correspondence analysis (DCA) of forest floor plants, *B* maximum DBH of trees, *C* the number of tree species with DBH ≥ 5 cm, *D* the number of floor plant species, *E* scores of the first axis in DCA of trees with DBH ≥ 5 cm

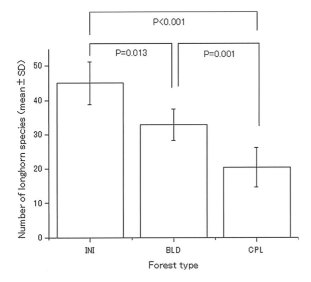

Fig. 3 Longicorn species richness of three forest types which were discriminated with CCA (Fig. 2). Results of ANOVA between the three types are given (corrected with Bonferroni adjustment)

leaved stands in Ogawa and the conifer (Japanese cedar) plantation stands in Satomi: the numbers of individuals and species were the largest in the youngest plots in both sites (O1, O4 and S3 in Table 1). One possible factor that may explain this is that the young plots abound with coarse wood debris (CWD), which is a suitable material for oviposition for some longicorn species (Linsley 1959, 1961). A preliminary CWD census in Ogawa showed that the youngest plots (O1 and O4) had

the largest amount of fallen logs or twigs (≥100 mm in length and ≥10 mm in diameter) among the ten plots studied (Makino et al., unpublished data). These dead wood materials were produced by the clear-cutting process, and they may have attracted longicorn adults seeking substrates for oviposition or adults may have emerged from them.

Another possibility is that the number of flowers that attract anthophilous longicorn adults is larger in young stands than in older ones. Because many longicorns visit flowers to feed on nectar and/or pollen (e.g. Linsley 1959, 1961; Kuboki 1987, 1998; Hanks 1999), their abundance may be positively correlated with the numbers of flowers present there. In addition, the composition of flowering species probably differs between young and old stands, and flowering species that are abundant in young plots may attract more longicorn species than the dominant flowering species of old plots, based on the observation that some anthophiolous longicorns show preferences for the flowers that they visit (e.g. Kuboki 1987). A monitoring study of the densities and compositions of both the flowering species and the longicorns visiting them throughout their active period would test these hypotheses.

Maeto and Makihara (1999) and Sayama et al. (2005) monitored cerambycids in broad-leaved stands of different ages in Ogawa by means of Malaise traps and odor attractant traps, respectively. These researchers found that the species richness was smallest in the youngest plots, which is contrary to what we found. The reason for this difference in results is unclear. Following the line of reasoning mentioned above, however, the amount of CWD or density and compositions of flowers may have been different between their plots and the ones established by us in the present investigation, particularly in terms of the young plots. Our youngest plots were 1–4 years old, while those of Maeto and Makihara (1999) and Sayama et al. (2005) were 5 years old after clear-cutting; although the difference in age is as little as 1 year and a maximum of 4 years, early changes in the resources for oviposition (CWD) or for nutrition (flowers) may be much greater than expected. A continuous monitoring of longicorns as well as CWD and flowers throughout the initial forest succession process, from 0 to 10 years after clear-cutting, for example, would be helpful in explaining the difference in the results between the studies.

Effects of conifer plantation on longicorn diversity

The CCA divided the plots into three forest types, INI, BLD, and CPL, respectively, corresponding to very young plots at an initial stand stage, secondary broad-leaved plots 12 to over 100 years old and conifer plantation plots over 8 years old (Fig. 2). The forest type INI comprised the youngest three plots of all (O1, O4 in Ogawa and S3 in Satomi), suggesting that very young conifer plantations are similar to grassland-like, open fields just after clear-cutting in terms of insect diversity.

Table 2 Longicorn species with significant index values (IndVals) for two of three forest types discriminated by canonical correspondence analysis (CCA) (Fig. 2), with the number of individuals collected in each forest type

Forest type[a]	Species[b]	IndVal	Number of individuals		
			INI	BLD	CPL
INI	*Dinoptera minuta*	80.0*	156	8	8
INI	*Leptura ochraceofasciata*	67.2**	73	29	29
INI	*Pidonia amentata*	94.1**	57	4	4
INI	*Chlorophus japonicus*	100.0**	63	0	0
INI	*Pareutetrapha simulans*	72.4*	25	3	3
INI	*Demonax transilis*	59.5*	11	7	7
INI	*Glenea relicta relicta*	67.9**	12	8	8
INI	*Xylotrechus grayii*	92.6**	16	3	3
INI	*Cyrtoclytus caproides*	97.8**	17	1	0
INI	*Prionus insularis*	85.2**	12	4	4
INI	*Leptura latipennis*	66.7*	18	0	0
INI	*Eumecoceara trivittata*	60.5*	11	3	0
INI	*Pterolophia angusta*	94.1**	12	2	0
INI	*Pterolophia zonata*	72.3**	6	5	1
INI	*Leptura modicenotata*	60.9*	9	0	2
INI	*Pogonocherus seminiveus*	60.2*	8	0	2
INI	*Phytoecia rufiventris*	66.7**	7	0	0
INI	*Megasemum quadricostulatum*	58.3*	3	0	0
INI	*Mesosa japonica*	56.1*	2	1	0
BLD	*Pidonia signifera*	67.5**	6	90	24
BLD	*Pterolophia tsurugiana*	70.5**	0	41	15
BLD	*Pidonia discoidalis*	90.4**	0	43	4
BLD	*Pidonia simillima*	67.6**	0	35	9
BLD	*Acalolepta sejuncta*	60.6*	1	24	7
BLD	*Judolidia japonica*	75.0*	0	21	0
BLD	*Parastrangalis shikokensis*	61.0*	0	15	3

Values significant at:
** $P < 0.01$; * $P < 0.05$
[a] INI, Initiation stage stands; BLD, broad-leaved stands
[b] No species had significant IndVal for the forest type CPL (coniferous plantations)

In this initial stage, only little divergence seems to occur between secondary broad-leaved stands and conifer plantations. The difference in longicorn assemblages and species richness between the broad-leaved and conifer plots increased thereafter.

The forest types INI and BLD both had longicorn species with significant IndVals; these species therefore preferentially occurred in younger or older broad-leaved plots, respectively. In INI, which is characterized as a stand at a very initial stage after clear-cutting or as a new plantation, there are many species with statistically significant IndVals. As mentioned earlier, those open and light plots immediately after clear-cutting may attract various longicorns which are searching for oviposition sites or for flowers as a food source. For example, the genera such as *Leptura* or *Judolia*, which have relatively large bodies, can fly long distances to visit flowers (Starzyk and Starzyk 1975); they may even have a tolerance for the higher temperatures and light intensities found in open fields (Kuboki 1987). The significant IndVals of three *Leptura* species (Table 2) are consistent with this reasoning: these anthophilous species appear to have been attracted to flower-rich resources in INI.

In contrast, no species had a significant IndVal for CPL, showing that no species preferentially or selectively occurred in conifer stands that were over 8 years old after plantation. This is well illustrated by Fig. 4, which shows that most of the species collected in the conifer plots were common to INI and/or BLD. Although four species were collected in the conifer stands only (Fig. 4), the catches were very small (one to four individuals per species: see Appendix). In addition, the

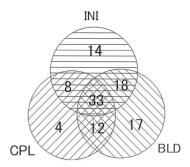

Fig. 4 The number of longicorn species collected in a single or multiple forest type. Three *circles* represent the number of species in the three forest types denoted as *INI* (initiation stage), *BLD* (broad-leaved stands) and *CPL* (conifer plantations). *Overlapped areas* show the number of species that are common to two or three forest types

recorded host plants of these species are mainly broad-leaved trees and do not include Japanese cedar (Kojima and Nakamura 1986). Therefore, these species most likely arrived quite by accident from outside of the conifer plots or emerged from understory trees in the plots. Makihara et al. (2004), who examined longicorn faunas of various parts of eastern Honshu, the mainland of Japan, also found that those species that occurred in Japanese cedar plantations were also collected in broad-leaved forests in the same area. This result also suggests that conifer plantations are incomplete substitutes of broad-leaved stands in terms of longicorn assemblages.

Japanese cedars are not attractive host plants for longicorns. Kojima and Nakamura (1986) published the host records of 468 Japanese longicorn species, of which 30 (6%) feed on (or emerge from) Japanese cedar. However, only two species, *Semanotus japonicus* and *Anaglyptus subfasciatus*, are recorded as specialists of Japanese cedar or related conifers; the other species usually have wider host ranges, including broad-leaved trees (Kojima and Nakamura 1986). These results suggest that a great majority of Japanese longicorn species can live without Japanese cedar, which dominate forested areas in many parts of Japan. Conversely, a significant decrease in the area occupied by broad-leaved forests and an increase in the area established by cedar plantations may lead to the impoverishment of longicorn faunas because most species simply have hosts other than Japanese cedar. Although conifer plantations often have rich understories of various herbs and trees (Ito et al. 2004; Nagaike et al. 2005; Tanaka et al., unpublished), our present results suggest that they are not large or diverse enough to carry a whole range of longicorns that would inhabit broad-leaved stands of comparable ages. Based on our results, it appears that those species with high IndVals for broad-leaved stands (BLD in Table 2) decrease in diversity and in population size with the conversion to a conifer plantation.

A number of other studies have also shown negative effects on the species richness of insects following the conversion of broad-leaved stands to monoculture plantations of conifers; these include carabid beetles in *Picea sitchensis* plantations (Fahy and Gormally 1998) and weevil beetles in *Larix kaempferi* (Ohsawa 2005). Additionally, the species compositions of various insects have been reported to differ between conifer plantations and broad-leaved forests, such as longicorn beetles (Shibata et al. 1996; Ohsawa 2004b; Maeto et al. 2002), clickbeetles (Ohsawa 2004a), cerambycid beetles (Day et al. 1993) and ants (Maeto and Sato 2004). It is apparent from these studies, including the present one, that conifer plantations cannot totally replace natural or semi-natural broad-leaved forests in terms of insect biodiversity. In order to conserve biodiversity at the landscape level, we should avoid an extensive monoculture plantation of conifers and attempt to retain natural broad-leaved stands within and among the plantations (Lindenmayer and Franklin 2002).

Implications of conifer plantation for ecosystem functioning

Decomposition is one of the major aspects of ecosystem functions. Although decomposition is principally brought about by microorganisms, it is often mediated by saproxylic insects (Grove 2002; Hartely and Jones 2004). Longicorn beetles may play an important role in promoting or accelerating decomposition processes through, for example, larval feeding on dead or moribund trees. In fact, one experiment revealed that the time required for wood to decay increased from 12 to 20 years in oak and from 7 to 12 years in pine when the macrofauna, including longicorns, was excluded from wood (Mamaev 1961, cited in Dajoz 2000).

From the viewpoint of ecosystem functions, longicorn adults may have an important function as potential pollinators (Maeto et al. 2002). Kato et al. (1990) found that longicorn beetles were the most frequent flower visitors among the beetles in a beech forest. Further, adults of the genus *Pidonia* often visit species-specific flowers (Kuboki 1987). Although their status as pollinator is not clear yet in almost all species, Kuboki (1998) recorded a case of pollination by *Pidonia* adults in *Magnolia sieboldii*.

An extensive conifer plantation may lead to the degradation of these functions through the impoverishment of local or regional longicorn faunas. Detailed studies focusing on quantitative analyses of ecosystem functions by insects are needed.

Acknowledgements We are grateful to Prof. T. Nakashizuka of Tohoku University for his helpful suggestions during the study. We also thank Mr. H. Makihara of Forestry and Forest Products Research Institute for identifying many longicorn specimens and providing us with useful information on the biology of longicorns. This work was partly funded by the Research Institute for Humanity and Nature and conducted as a part of its research project "Sustainability and Biodiversity Assessment on Forest Utilization Options," and by the grant "Development of eco-friendly management technology of water and agro-forested-aqua-ecosystem in watershed and estuary areas" from the Ministry of Agriculture, Forestry and Fisheries, Japan.

Appendix

Table 3 Number of individuals of longicorn beetles collected in the study plots[a]

Number	Species	Study plot																		Total number
		O1	O4	O12	O24	O51	O54	O71	O128	O174	O178	S3	S7	S9	S20	S29	S31	S75	S76	
1	*Corymbia succedanea*	27	0	0	1	0	0	0	1	0	0	199	43	4	2	4	3	0	0	284
2	*Parastrangalis nymphula*	5	9	35	12	13	9	26	8	17	27	11	25	4	4	12	23	7	4	251
3	*Dinoptera minuta*	29	117	11	21	46	2	10	0	2	3	10	2	3	0	0	0	3	0	259
4	*Leptura ochraceofasciata*	19	24	4	9	13	10	7	0	10	9	30	11	2	0	6	0	6	4	164
5	*Pidonia signifera*	1	4	8	10	12	12	16	1	9	22	1	3	1	2	4	3	9	2	120
6	*Stenygrium quadrinotatum*	5	20	3	18	0	0	0	12	10	0	1	0	1	0	0	0	0	0	70

Table 3 (Contd.)

Number	Species	O1	O4	O12	O24	O51	O54	O71	O128	O174	O178	S3	S7	S9	S20	S29	S31	S75	S76	Total number
7	*Acalolepta fraudatrix*	6	6	3	4	4	12	4	1	5	7	7	0	0	0	0	1	4	1	65
8	*Pidonia amentata amentata*	6	39	1	1	0	0	3	0	0	0	12	3	1	0	0	0	0	0	66
9	*Chlorophus japonicus*	47	4	0	0	0	0	0	0	0	0	12	0	0	0	0	0	0	0	63
10	*Pterolophia tsurugiana*	0	0	2	3	2	2	6	7	11	8	0	0	2	2	0	8	3	0	56
11	*Pidonia puziloi*	1	1	4	1	2	4	7	0	3	6	0	1	1	4	6	4	1	0	46
12	*Pareutetrapha simulans*	2	22	4	10	2	2	3	1	0	0	1	0	1	0	0	0	2	0	50
13	*Pidonia discoidalis*	0	0	1	1	5	18	9	1	4	4	0	0	0	0	3	0	1	0	47
14	*Pidonia simillima*	0	0	1	0	7	8	3	7	6	3	0	0	0	2	3	1	3	0	44
15	*Pterolophia jugosa*	8	2	4	0	5	2	0	8	3	1	4	0	4	0	1	2	0	0	44
16	*Distenia gracliis*	21	0	5	1	1	5	2	0	2	3	1	0	2	0	0	2	0	0	45
17	*Pterolophia granulata*	0	2	0	0	1	0	1	0	0	0	7	26	2	0	1	1	0	0	41
18	*Asaperda agapanthina*	1	7	0	1	1	0	2	0	0	6	2	3	3	3	2	6	0	2	39
19	*Pseudocalamobius japonicus*	0	0	1	12	0	0	0	0	0	0	0	3	1	1	7	0	6	2	33
20	*Acalolepta sejuncta*	1	0	8	3	2	3	0	5	1	2	0	0	0	0	0	5	2	0	32
21	*Demonax transilis*	1	5	4	2	1	0	1	0	1	3	5	2	0	1	2	0	0	2	30
22	*Nupserha marginella*	1	11	2	4	0	1	0	3	2	0	1	2	0	0	0	0	2	2	31
23	*Xylotrechus emaciatus*	1	0	6	0	3	2	0	0	0	5	1	0	1	2	1	5	2	1	30
24	*Clytus melaenus*	25	0	0	0	0	0	1	1	0	0	0	0	0	0	0	0	0	0	27
25	*Glenea relicta*	2	3	1	2	1	0	2	0	0	0	7	2	5	0	0	1	0	0	26
26	*Lemula rufithorax*	0	0	0	4	1	0	0	20	0	1	0	0	0	0	0	0	0	0	26
27	*Judolidia bangi*	2	0	3	0	1	0	4	0	1	2	1	2	1	1	1	3	1	0	23
28	*Xylotrechus cuneipennis*	5	4	0	3	2	0	1	2	0	3	0	0	1	0	0	0	0	0	21
29	*Pidonia aegrota*	2	1	0	2	3	1	6	2	1	0	1	0	0	0	0	1	0	1	21
30	*Judolidia japonica*	0	0	5	2	3	2	5	0	4	0	0	0	0	0	0	0	0	0	21
31	*Xylotrechus grayii*	4	6	0	0	0	0	0	0	0	0	6	3	0	0	0	0	0	0	19
32	*Pidonia grallatrix*	0	1	0	0	0	0	15	0	0	2	0	0	0	0	0	0	0	0	18
33	*Cyrtoclytus caproides*	4	4	0	0	0	0	1	0	0	0	9	0	0	0	0	0	0	0	18
34	*Parastrangalis shikokensis*	0	0	2	0	0	1	5	1	1	5	0	0	0	0	0	3	0	0	18
35	*Prionus insularis*	3	3	0	0	0	0	0	0	0	1	6	1	2	1	0	0	0	0	17
36	*Leptura latipennis*	17	1	0	0	0	0	0	0	0	0	0	0	0	0	0	0	0	0	18
37	*Lemula decipiens*	0	3	1	3	1	0	0	9	0	0	0	0	0	0	0	0	0	0	17
38	*Graphidessa venata*	0	0	0	0	0	0	0	1	0	4	0	0	0	0	0	6	2	0	13
39	*Pterolophia caudata*	0	0	5	3	6	0	0	0	0	0	1	0	0	0	1	0	0	0	16
40	*Psephactus remiger*	2	1	2	0	1	0	2	0	5	1	0	0	0	0	0	0	1	0	15
41	*Sybra subfasciata*	7	0	1	1	1	0	1	0	2	0	1	0	0	0	0	0	0	0	14
42	*Eumecoceara trivittata*	8	3	0	0	0	1	1	0	0	1	0	0	0	0	0	0	0	0	14
43	*Pterolophia angusta*	10	1	0	0	0	0	0	2	0	0	1	0	0	0	0	0	0	0	14
44	*Pterolophia zonata*	2	3	1	2	2	0	0	0	0	0	1	1	0	0	0	0	0	0	12
45	*Exocentrus testudineus*	2	0	2	1	0	0	0	0	0	0	2	0	0	1	0	3	1	0	12
46	*Rhaphuma xenisca*	7	0	1	1	0	0	0	0	0	0	0	0	2	0	0	0	0	1	12
47	*Menesia sulphurata*	0	2	3	1	0	0	1	0	0	1	1	0	0	0	2	0	0	0	11
48	*Leptura modicenotata*	3	6	0	0	0	0	0	0	0	0	0	1	1	0	0	0	0	0	11
49	*Oberea infranigrescens*	0	3	0	0	0	0	0	1	0	0	1	3	2	0	0	1	0	0	11
50	*Mesosella simiola*	1	0	0	2	0	0	0	2	0	3	1	0	0	0	0	0	0	0	9
51	*Phymatodes albicinctus*	0	0	2	2	0	0	1	0	0	0	0	0	4	0	0	0	0	0	9
52	*Pogonocherus seminiveus*	4	4	0	0	0	0	0	0	0	0	0	0	2	0	0	0	0	0	10
53	*Callidiellum rufipenne*	0	0	0	0	0	0	0	0	0	0	3	0	0	0	0	4	0	2	9
54	*Japonostrangalia dentatipennis*	0	0	1	0	1	2	0	1	0	1	1	0	0	0	0	1	0	0	8
55	*Phytoecia rufiventris*	0	5	0	0	0	0	0	0	0	0	2	0	0	0	0	0	0	0	7
56	*Clytus auripilis*	6	0	0	0	0	0	0	1	0	0	0	0	0	0	0	0	0	0	7
57	*Anastrangalis scotodes*	5	0	0	0	0	0	0	0	0	0	0	0	0	0	2	0	0	0	7
58	*Monochamus subfasciatus*	0	2	0	0	0	0	0	5	0	0	0	0	0	0	0	0	0	0	7
59	*Leiopus stillatus*	0	0	0	0	0	1	1	0	1	3	0	0	0	0	0	0	0	0	6
60	*Palimna liturata*	0	0	0	0	0	0	0	0	0	6	0	0	0	0	0	0	0	0	6
61	*Gaurotes doris*	2	1	0	0	0	0	1	0	0	0	0	0	0	0	0	1	0	0	5
62	*Plagionotus christophi*	5	0	0	0	0	0	0	0	0	0	0	0	0	0	0	0	0	0	5
63	*Eumecocera gleneoides*	0	0	1	0	0	3	0	0	0	0	0	0	1	0	0	0	0	0	5
64	*Phymatodes testaceus*	5	0	0	0	0	0	0	0	0	0	0	0	0	0	0	0	0	0	5
65	*Rhopaloscelis masculatus*	0	0	0	1	0	0	0	4	0	0	0	0	0	0	0	0	0	0	5
66	*Toxotinus reini*	0	0	0	0	0	1	0	0	0	0	0	1	0	1	2	0	0	0	5
67	*Pidonia chairo*	0	0	0	0	0	0	4	0	1	0	0	0	0	0	0	0	0	0	5
68	*Uraecha bimaculata*	0	0	0	0	1	0	0	2	0	1	0	0	0	0	0	1	0	0	5
69	*Megasemum quadricostulatum*	2	0	0	0	0	0	0	0	0	0	1	0	0	0	0	0	1	0	4
70	*Miccolamia cleroides*	0	0	0	0	0	0	0	0	0	0	0	0	0	1	3	0	0	0	4
71	*Falsomesosella gracilior*	0	1	0	1	0	0	0	1	0	1	0	0	0	0	0	0	0	0	4

Table 3 (Contd.)

Number	Species	O1	O4	O12	O24	O51	O54	O71	O128	O174	O178	S3	S7	S9	S20	S29	S31	S75	S76	Total number
72	*Rondibilis saperdina*	0	0	4	0	0	0	0	0	0	0	0	0	0	0	0	0	0	0	4
73	*Arhopaloscelis bifasciatus*	0	0	0	0	0	0	0	3	0	1	0	0	0	0	0	0	0	0	4
74	*Anoploderomorpha excavata*	1	0	1	1	0	0	0	0	0	0	1	0	0	0	0	0	0	0	4
75	*Macroleptura regalis*	0	0	0	0	0	0	0	0	0	0	2	1	0	0	0	0	0	0	3
76	*Paramenesia kasugensis*	0	1	0	0	0	0	0	2	0	0	0	0	0	0	0	0	0	0	3
77	*Anoplophora malasiaca*	0	0	1	0	0	0	0	1	0	1	0	0	0	0	0	0	0	0	3
78	*Mesosa japonica*	0	1	0	1	0	0	0	0	0	0	1	0	0	0	0	0	0	0	3
79	*Paraclytus excultus*	0	2	0	0	0	1	0	0	0	0	0	0	0	0	0	0	0	0	3
80	*Pseudalosterna misella*	0	0	0	0	0	0	0	0	1	0	1	0	0	0	0	1	0	0	3
81	*Cleptometopus bimaculata*	0	0	0	1	0	0	0	0	1	0	0	1	0	0	0	0	0	0	3
82	*Eutetrapha chrysochloris*	0	0	0	0	0	2	0	1	0	0	0	0	0	0	0	0	0	0	3
83	*Pterolophia leiopoina*	0	1	1	0	0	1	0	0	0	0	0	0	0	0	0	0	0	0	3
84	*Phymatodes maaki*	0	0	0	1	0	0	0	0	0	0	0	1	0	0	0	0	0	0	2
85	*Brachyclytus singularis*	2	0	0	0	0	0	0	0	0	0	0	0	0	0	0	0	0	0	2
86	*Eustrangalia distenioides*	0	1	0	0	0	0	0	0	0	1	0	0	0	0	0	0	0	0	2
87	*Mesosa senilis*	2	0	0	0	0	0	0	0	0	0	0	0	0	0	0	0	0	0	2
88	*Xenicotela pardalina*	0	1	0	0	0	0	0	0	0	0	0	0	0	1	0	0	0	0	2
89	*Mesosa longipennis*	2	0	0	0	0	0	0	0	0	0	0	0	0	0	0	0	0	0	2
90	*Epiclytus yokoyamai*	0	1	0	0	0	0	0	1	0	0	0	0	0	0	0	0	0	0	2
91	*Lemula nishimurai*	0	0	0	1	0	0	0	0	0	0	0	0	0	0	0	0	0	0	1
92	*Exocentrus lineatus*	1	0	0	0	0	0	0	0	0	0	0	0	0	0	0	0	0	0	1
93	*Mesosa hirsuta*	0	0	0	0	0	0	0	0	0	0	1	0	0	0	0	0	0	0	1
94	*Grammographus notabilis*	1	0	0	0	0	0	0	0	0	0	0	0	0	0	0	0	0	0	1
95	*Asaperda rufipes*	0	0	0	0	0	0	0	0	1	0	0	0	0	0	0	0	0	0	1
96	*Nanohammus rufescens*	0	0	0	0	0	0	0	1	0	0	0	0	0	0	0	0	0	0	1
97	*Pterolophia castaneivora*	1	0	0	0	0	0	0	0	0	0	0	0	0	0	0	0	0	0	1
98	*Atimura japonica*	1	0	0	0	0	0	0	0	0	0	0	0	0	0	0	0	0	0	1
99	*Stehomalus takaosanus*	0	0	0	0	0	0	0	0	0	1	0	0	0	0	0	0	0	0	1
100	*Egesina bifasciana*	0	0	0	0	0	0	0	0	0	0	0	0	0	0	0	1	0	0	1
101	*Eumecocera argyrosticta*	0	0	0	0	0	0	0	0	0	0	0	1	0	0	0	0	0	0	1
102	*Rhopaloscelis unifasciatus*	0	0	0	0	0	0	0	0	0	0	0	0	1	0	0	0	0	0	1
103	*Rhuphum diminuta*	0	0	0	0	0	0	1	0	0	0	0	0	0	0	0	0	0	0	1
104	*Nakanea vicaria*	1	0	0	0	0	0	0	0	0	0	0	0	0	0	0	0	0	0	1
105	*Mesosa poecila*	0	0	0	0	0	1	0	0	0	0	0	0	0	0	0	0	0	0	1
106	*Dolichoprosopus yokoyamai*	0	0	0	0	0	0	0	0	1	0	0	0	0	0	0	0	0	0	1

[a] See Section Study areas and Table 1 for explanation of the plots

References

Buddle CM, Spece JR, Longor DW (2000) Succession of boreal forest spider assemblages following wildfire and harvesting. Ecography 23:424–436

Dajoz R (2000) Insects and forests. Intercept, London

Day KR, Marshall S, Heaney C (1993) Associations between forest type and invertebrate: ground beetle community patterns in a natural oakwood and juxtaposed conifer plantations. Forestry 66:37–50

Dufrêne M, Legendre P (1997) Species assemblages and indicator species: the need for a flexible asymmetrical approach. Ecol Monogr 67:345–366

Fahy O, Gormally M (1998) A comparison of plant and carabid beetle communities in an Irish oak woodland with a nearby conifer plantation and clearfelled site. For Ecol Manage 110:263–273

Grove SJ (2002) Saproxylic insect ecology and the sustainable management of forests. Annu Rev Ecol Syst 23:1–23

Hanks LM (1999) Influence of the larval host plant on reproductive strategies of cerambycid beetles. Annu Rev Entomol 4:483–505

Hartely SE, Jones TH (2004) Insect herbivores, nutrient cycling and plant productivity. In: Weisser WW, Siemann E (eds) Insects and ecosystem function. Springer, Berlin, pp 27–52

Hartley MJ (2002) Rationale and methods for conserving biodiversity in plantation forests. For Ecol Manage 155:81–95

Inoue T (2003) Chronosequential change in a butterfly community after clear-cutting of deciduous forests in a cool temperate region of central Japan. Entomol Sci 6:151–163

Ito S, Nakayama R, Buckley GP (2004) Effects of previous land-use on plant species diversity in semi-natural and plantation forests in a warm-temperate region in southeastern Kyusyu, Japan. For Ecol Manage 196:213–225

Japan Agriculture Statistics Association (2001) Analysis of structure and forest management in Japanese forestry based on forestry census 2000 (in Japanese). Japan Agriculture Statistics Association, Tokyo

Kato M, Kakutani T, Inoue T, Itino T (1990) Insect–flower relationship in the primary beech forest of Ashu, Kyoto: an overview of the flowering phenology and the seasonal pattern of insect visits. Contr Biol Lab Kyoto Univ 27:309–375

Kojima K, Nakamura S (1986) Host tree records of Japanese longicorn beetles. Hiba-Kagaku Kyoiku Shinkosha, Shobara

Kuboki M (1987) Logicorn beetles of the genus *Pidonia* in Japan (in Japanese). Bunichi Sogo Shuppan Press, Tokyo

Kuboki M (1998) Ceramycids. In: Hidaka et al. (eds) Encyclopedia of animals in Japan, vol. 10. Insects III (in Japanese). Heibonsha, Tokyo, pp 141–145

Lindenmayer DB, Franklin JF (2002) Conserving forest biodiversity. A comprehensive multiscaled approach. Island Press, Washington D.C.

Linsley EG (1959) Ecology of Cerambycidae. Annu Rev Entomol 4:99–138

Linsley EG (1961) The Cerambycidae of North America, vol. 18. Part I. Introduction. University of California Publications onEntomology. University California Press, Berkeley

Maeto K, Makihara H (1999) Changes in insect assemblages with secondary succession of temperate deciduous forests after clear-cutting (in Japanese with English summary). Jpn J Entomol (NS) 2:11–26

Maeto K, Sato S (2004) Impacts of forestry on ant species richness and composition in warm-temperate forests of Japan. For Ecol Manage 187:213–223

Maeto K, Sato S, Miyata H (2002) Species diversity of longicorn beetles in humid warm-temperate forests: the impact of forest management practices on old-growth forest species in southwestern Japan. Biodiv Conserv 11:1919–1937

Magurran AE (2004) Measuring biological diversity. Blackwell, Malden

Makihara H, Koma Y, Ikeda S, Goto T (2004) Investigations of longicorn beetles (Coleoptera, Cerambycidae) indicator in Satoyama (II) – Sugi stand in Nanakai Village of Ibaraki Pref. (in Japanese). Trans Jpn For Soc Kanto Branch 55:217–220

Makino S, Goto T, Sueyoshi M, Okabe K, Hasegawa M, Hamaguchi K, Tanaka H, Okochi I (2006) The monitoring of insects to maintain biodiversity in Ogawa forest reserve. Environ Monit Assess 120: 477–485. doi: 10.1007/s10661-005-9074-810.1007/s10661-005-9074-8

Mamaev BM (1961) Activity of larger invertebrates as one of the main factors of natural destruction of wood (in Russian). Pedobiologia 1:38–52

McCune, B., Mefford, M. J. (1999) PC-ORD for Windows. Multivariate analysis of ecological data, version 4.20. MJM Software Design, Gleneden Beach

Mizoguchi Y, Morisawa T, Ohtani Y (2002) Climate of Ogawa forest reserve. In: Nakashizuka T, Matsumoto Y (eds) 2002. Diversity and interaction in a temperate forest community: Ogawa forest reserve of Japan. Springer, Tokyo, pp 11–18

Mizota K, Imasaka S (1997) Comparison of flower-visiting beetle communitites between natural and artificial forests in southern Kii Peninsula: use of benzyl acetate traps. Res Bull Hokkaido Univ For 54:299–326

Moore SE, Allen HL (1999) Plantation forestry. In: Malcom L, Hunter JR (eds) Maintaining biodiversity in forest ecosystems, Cambridge University Press, Cambridge, pp 400–433

Nagaike T, Kamitani T, Nakashizuka T (2005) Effects of different forest management systems on plant species diversity in a *Fagus renata* forested landscape of central Japan. Can J For Res 35:2832–2840

Ohashi H, Nohira T, Watanabe K (1992) Insects collected with a flower odor attractant (in Japanese). Bull Gifu For Inst 20:15–48

Ohsawa M (2004a) Comparison of Elaterid biodiversity among larch plantations, secondary forests, and primary forests in the central mountainous region in Japan. Ann Entomol Soc Am 97:770–774

Ohsawa M (2004b) Species richness of Cerambycidae in larch plantations and natural broad-leaved forests of the central mountainous region of Japan. For Ecol Manage 189:375–385

Ohsawa M (2005) Species richness and composition of Curculionidae (Coleoptera) in a conifer plantation, secondary forest, and old-growth forest in the central mountainous region of Japan. Ecol Res 20:632–645

Oliver CD (1981) Forest development in North America following major disturbances. For Ecol Manage 3:153–168

Palik B, Engstrom T (1999) Species composition. In: Malcom L, Hunter JR (eds) Maintaining biodiversity in forest ecosystems. Cambridge University Press, Cambridge, pp 65–94

Sayama K, Makihara H, Inoue T, Okochi I (2005) Monitoring longicorn beetles in different forest types using collision traps baited with chemical attractants. Bull FFPRI 4:189–199

Shibata E, Sato S, Sakuratani Y, Sugimoto T, Kimura F, Ito F (1996) Cerambycid beetles (Coleoptera) lured to chemicals in forests of Nara Prefecture, central Japan. Ann Entomol Soc Am 89:835–842

SPSS Inc (1998) SYSTAT for Windows, version 9. SPSS Inc, Chicago, Ill.

Starzyk JR, Starzyk K (1975) Preliminary investigation on the penetration in search of food of anthophilous species of the Cerambycidae family (Coleoptera) in the Tatra National Park. Acta Agrar Selvestria 15:93–110

StatSoft Inc. (2000) STATISTICA for Windows. StatSoft Inc., Tulsa, Okla.

Sueyoshi M, Maetô K; Makihara H, Makino S, Iwai T (2003) Changes in dipteran assemblages with secondary succession of temperate deciduous forests following clear-cutting (in Japanese with English abstract). Bull FFPRI 2:171–192

Suzuki W (2002) Forest vegetation in and around Ogawa Forest Reserve in relation to human impact. In: Nakashizuka T, Matsumoto Y (eds) Diversity and interaction in a temperate forest community: Ogawa forest reserve of Japan. Springer, Tokyo, pp 27–41

ter Braak CJF, Smilauer P (2002) CANOCO reference manual and CANODRAW for Windows User's Guide: software for canonical community ordination (version 4.5). Microcomputer Power, Ithaca

Trofymow JA, Addison J, Blackwell BA, He F, Preston CA, Marshall VG (2003) Attributes and indicators of old-growth and successional Douglas-fir forests on Vancouver Island. Environ Rev 11:S1–S18

Ecol Res (2007) 22: 382–389
DOI 10.1007/s11284-007-0360-5

Yuji Isagi · Ryunosuke Tateno · Yu Matsuki
Akira Hirao · Sonoko Watanabe · Mitsue Shibata

Genetic and reproductive consequences of forest fragmentation for populations of *Magnolia obovata*

Received: 27 February 2006 / Accepted: 17 September 2006 / Published online: 16 March 2007
© The Ecological Society of Japan 2007

Abstract In order to evaluate the consequences of forest fragmentation on populations of *Magnolia obovata*, we compared genetic diversity and reproductive characteristics at two nearby sites, one conserved and one fragmented. The genetic diversity between adults trees of the different sites was not significantly different. However, saplings in the conserved site showed a significantly higher genetic diversity than both adult trees in the conserved site and saplings in the fragmented sites; this was found to be the result of the larger gene flow into the conserved site. The density of the adult trees was significantly related to all of the reproductive traits analyzed (fertilization of ovules, insect attack to seeds, ovules that developed into seeds and outcrossing at the stage of seeds) at both sites. At both sites, fertilization of ovules and insect attack on seeds were positively correlated to adult tree density while outcrossing rate was negatively correlated to adult tree density. The fertilization of ovules and outcrossing were more dependent on adult tree density in the fragmented site than in the conserved site. The probability of ovules developing into outcrossed seeds showed a negative correlation with adult tree density at both sites, indicating the advantage of low density for this species and possibly implying a resilience to habitat fragmentation. A two-generation-analysis did not identify significant differences between sites in terms of the structure of the pollen pool and the number of pollen donors. Although fragmentation affected reproductive characteristics, the effect on seedling establishment and subsequent survival remains to be determined. Proposals for future studies that will assist in the development of management strategies for forests suffering fragmentation are made.

Keywords Fecundity · Forest fragmentation · Fruition · Regeneration · Pollination

Y. Isagi (✉)
Graduate School of Agriculture, Kyoto University,
Kitashivakawa Oiwake-cho, Sakyo-ku, Kyoto 606-8502, Japan
E-mail: isagiy@kais.kyoto-u.ac.jp
Tel.: +81-75-7536422
Fax: +81-75-7536129

R. Tateno
Graduate School of Agriculture, Kagoshima University,
Korimoto 1-21-24, Kagoshima 890-0065, Japan

Y. Matsuki · S. Watanabe
Graduate School for International Development and
Cooperation, Hiroshima University, Kagamiyama 1-5-1,
Higashi-Hiroshima 739-8529, Japan

A. Hirao
Graduate School of Environmental Earth Science,
Hokkaido University, Sapporo 060-0810, Japan

M. Shibata
Tohoku Research Center,
Forestry and Forest Products Research Institute,
Morioka, Iwate 020-0123, Japan

Introduction

Forest fragmentation as a result of the extensive anthropogenic use of resources is a threat to biodiversity that may have an influence on forest ecosystems and cause detrimental effects, both genetically as well as ecologically. The level of genetic diversity may decrease in direct response to shrinking populations of trees, followed by further genetic erosion through increases in random genetic drift, inbreeding, and reduced gene flow (Young et al. 1996; Oostermeijer et al. 2003). Fragmented populations may suffer inbreeding depression, increased susceptibility to diseases and pests, the fixation of deleterious alleles, and the loss of self-incompatibility alleles. Within a short period of time, inbreeding may lead to a population with a reduced fitness and lower viability. Reduced genetic diversity may also limit the ability of the population to adapt to environmental changes. Fragmentation may also have ecological consequences for forest ecosystems. For example, changes in the activity, abundance, and species of pollinators and increased distances between trees and populations may result in pollen limitation and lower seed set. Shifts in interactions between species following fragmentation may also alter the characteristics of plant reproduction.

Fragmentation negatively affects the reproductive success of plants by reducing performance, ranging from the activity of pollinators (Aizen and Feinsinger 1994a, b; Quesada et al. 2003), pollen deposition (Cunningham 2000; Cascante et al. 2002; Quesada et al. 2003) and seed set (Ghazoul et al. 1998; Fuchs et al. 2003; Quesada et al. 2004) to the regeneration of populations (Benitez-Malvido 1998; Cascante et al. 2002; Benitez-Malvido and Martínez-Ramos 2003). However, several studies have shown that forest fragmentation can have positive (Dick 2001; White et al. 2002; Dick et al. 2003) or neutral (Cascante et al. 2002; Fuchs et al. 2003) effects on a number of traits associated with plant reproduction.

Although the influence of forest fragmentation or deforestation on ecosystems and communities has been analyzed (see studies mentioned above), a relatively smaller number of studies have evaluated the genetic consequences of fragmentation (Aldrich and Hamrick 1998; Aldrich et al. 1998; White et al. 1999; Dayanandan et al. 1999; Collevatti et al. 2001; Dick 2001; Obayashi et al. 2002; White et al. 2002; Fuchs et al. 2003), and the locations of the study sites are biased toward tropical regions (Lowe et al. 2005). These studies, however, have revealed that the effects of habitat fragmentation on the genetic traits of tree populations are more complex than expected (Aldrich and Hamrick 1998; Dick 2001; White et al. 2002; Dick et al. 2003; Lowe et al. 2005).

The objectives of the present research were to assess the consequences of forest fragmentation on the genetic diversity, reproductive characteristics and gene flow of *Magnolia obovata* by comparing fragmented and non-fragmented sites located in a research area of the temperate region where this type of study has only rarely been conducted. The assessment includes an evaluation of the biological conservation value of remnant forest and establishes strategies for sustainable management based on maintaining or even restoring ecological function and genetic resources. In order to assess the consequences of forest fragmentation on the genetic and reproductive traits of *M. obovata*, we compared genetic diversity, distance of pollen movement, and rates of fertilization of ovules, insect attack to seeds, ovules that developed into seeds and outcrossing at the seed stage between conserved and fragmented sites.

Materials and methods

Study species

Magnolia obovata is a common, deciduous tree species growing to 30 m in height in Japanese temperate forests. The standing density of adults of this species is relatively low – a few trees per hectare. Flowers of *Magnolia* species do not secrete nectar and are primarily pollinated by beetles (Thien 1974; Judd et al. 2002), which are relatively inefficient as pollen vectors (Thien 1974). However, a variety of insects have been observed to visit the flowers of *M. obovata* (Tanaka and Yahara 1988). The flowers are protogynous, with petals closing between the female and male periods (Kikuzawa and Mizui 1990). Although individual flowers last for 3–4 days, a tree flowers for up to 40 days (Kikuzawa and Mizui 1990). The pistil contains 70–150 carpels, and each carpel can bear two seeds. Seeds reach maturity in the autumn and are dispersed by birds.

Study area

We compared conserved and fragmented forest sites in the Ogawa Forest Reserve and its surrounding area (36°56′N, 140°35′E), Ibaraki Prefecture, central Japan (Fig. 1). The sites are situated at an altitude of 610–660 m a.s.l. on a quasiplain in the southern part of the Abukuma Mountains. Mean annual air temperature and precipitation at a meteorological station in Ogawa (36°54′N, 140°35′E) during the study period were 10.7°C and 1910 mm, respectively (Moriguchi et al. 2002). The area is covered by a deciduous broad-leaved forest, and the dominant woody species in the canopy are *Quercus serrata*, *Fagus japonica*, and *F. crenata*. Intensive studies of the structure and dynamics of the plant community (Nakashizuka et al. 1992) and on the population dynamics of *Carpinus* (Shibata and Nakashizuka 1995), *Acer* (Tanaka 1995) and *Cornus* (Masaki et al. 1994) have been conducted in the area. The conserved site comprises 98 ha. The fragmented site has an elongated shape (approx. 1500×200 m long; 29 ha), is surrounded by a conifer plantation, paddy fields, golf courses, and pasture and is located 500 m to the northeast of the conserved site. The location and size (diameter at breast height; DBH) of all reproductive-age *M. obovata* were determined at both sites (Fig. 1).

Evaluation of genetic diversity

Trees with a DBH of more than 30 cm ($n = 88$ and 29 in the conserved and fragmented sites, respectively), and saplings ($n = 128$ and 92 in the conserved and fragmented sites, respectively) were located and genotyped using 11 microsatellite markers developed for *M. obovata* (Isagi et al. 1999). The height of the saplings ranged from 2.8 to 615 cm, with an average of 70.3 cm. The effective number of alleles (a_e) was calculated for trees and saplings at each site using the formula $a_e = 1/\Sigma x_i^2$), where x_i is the frequency of the ith allele for each locus (Hedrick 2005). The significance of the difference between the effective number of alleles of adults and saplings and those in the conserved and fragmented sites were compared using paired t-tests.

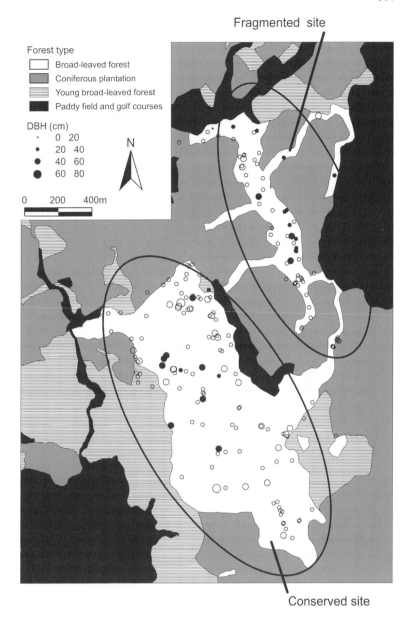

Fig. 1 Size and location of reproductive *Magnolia obovata* trees at the conserved and fragmented sites. *Filled circle* Adult tree from which fruit were collected

Fragmented site

Forest type
- Broad-leaved forest
- Coniferous plantation
- Young broad-leaved forest
- Paddy field and golf courses

DBH (cm)
- 0　20
- 20　40
- 40　60
- 60　80

N

0　　200　　400m

Conserved site

Pollination, seed set, and self-pollination rates

For the analysis of pollination and seed set, 119 and 136 mature fruit were collected in September 2003 from 14 and 14 adult trees in the conserved and fragmented sites, respectively (Fig. 1). The number of ovules per fruit was estimated by doubling the number of carpels since each carpel of *M. obovata* contains two ovules. The number of fertilized ovules was estimated by summing up the numbers of rotten seeds, seeds attacked by insects, and sound seeds. The germination of the seeds of the 255 fruit from the 28 trees was induced by a continuous temperature fluctuation cycle consisting of 12:12 h at 24°:16°C. The DNA of 414 germinated seedlings was extracted from the root tips (about 5 mm in length), and genotype analyses at microsatellite loci M6D3, M6D8, M10D3 and M10D8 were conducted to identify the pollination mode (self-pollinated or allogamous).

The relationship between conspecific density and rates of fertilization of ovules, insect attack to seeds, ovules that developed into seeds, and outcrossing at the stage of seeds were evaluated by fitting binomial generalized linear models (GLMs) (no fertilization = 0, fertilization = 1; no insect attack = 0, insect attack = 1; no development of ovule to a seed = 0, development of ovule to a seed = 1; self-pollination = 0, outcrossing = 1) with the logistic link using the R software package (ver. 2.2.1; R Development Core Team 2005). When self-pollination was included in the analysis, the average distance of pollen movement was 20.9 m (Isagi et al. 2004), and 77% of pollen exchange occurred a radius of 200 m from an adult tree. Based on this result, we evaluated the density of conspecifics around a mother tree by the number of trees within a radius of 200 m. The goodness-of-fit of the models was employed using the deviance chi-square test for the null hypothesis that the parameter estimates do not differ from zero.

Gene flow estimation by the Two-Generation analysis

Gene flow has been quantified by direct and indirect methods. Indirect methods (e.g. Wright 1951; Slatkin 1985) estimate the amount of gene flow per generation based on the contemporary genetic structure and diversity among populations, as this reflects pollen and seed movements over generations. Direct methods based on parentage analyses with highly informative genetic markers, such as microsatellites, can measure current rates of gene flow although such methods require that the genotypes of all offspring and possible parents within the research sites be determined. Where there are large number of possible parents, exclusion of unrelated parents can become ambiguous even with high exclusion probabilities. Hence, direct approaches are only feasible for populations with a relatively small number of individuals. At the conserved and fragmented sites 114 and 70 trees flowered in 2003, respectively, a number that would have rendered direct approaches laborious.

In order to evaluate rates of pollen movement, we employed a two-generation analysis (TwoGener; Smouse et al. 2001). In the TwoGener analysis, genotypes of maternal plants and their offspring are evaluated. Using four microsatellite loci, we determined the genotypes of outcrossed seedlings, of which 53 were from seven seed parents in the conserved site and 88 were from five seed parents in the fragmented site. The TwoGener analysis uses a statistic Φ_{ft}, which is an analog of the F_{ST} coefficient of genetic differentiation among populations (Wright 1951). When the F_{IS} value is more than zero because of a genetic structure within a population, the value Φ_{ft} might be overestimated. Therefore, the value Φ_{ft} was divided by $1 + F_{IS}$ (Austerlitz and Smouse 2001). F_{IS} was calculated with the FSTAT computer program (Goudet 1995). The Φ_{ft} value, which ranges from 0 to 1, represents the genetic differentiation among pollen pools sampled from

different maternal plants. Given the absence of local genetic structure among adult trees, an appropriate distribution of pollen dispersal and equal fecundity of the adult trees, we can estimate the effective number of pollen donors (N_{ep}) with the equation, $N_{ep} = 1/(2\Phi_{ft})$.

Results

Genetic diversity

The effective number of alleles was smaller at the fragmented site than at the conserved one for both adult trees and saplings (Table 1), but the difference was only significant for the saplings ($P = 0.014$). At the conserved site, there were significantly more effective alleles in the saplings than in the adults ($P = 0.0028$).

Pollination, seed set and germination

The fragmented site showed significantly higher ratios than the conserved site in terms of proportions of fertilized ovules, seeds attacked by insects, and ovules that developed into seeds (Table 2). The ratio of outcrossing at the seeds stage was higher at the fragmented site than the conserved site (Table 2), but the difference was not significant ($P = 0.09$).

Adult tree density within 200 m of the seed parents was significantly related to all reproductive traits analyzed at both sites (Fig. 2). The fertilization of ovules and insect attack to seeds were positively related to adult tree density, while outcrossing at the stage of seeds was negatively related to adult tree density (Fig. 2) at both sites. The ratios of ovules that developed into seeds responded oppositely to adult tree density at each site (Fig. 2c). The gradient of the regression curves for the fertilization of ovules and outcrossing were steeper at the fragmented site (Fig. 2a, d), indicating the larger dependence of these traits to adult tree density in the fragmented site.

Pollen pool structure and number of pollen donors

The estimated values of Φ_{FT} were 0.137 and 0.144 for the conserved and fragmented sites, respectively. As

Table 1 Effective number of alleles at 11 microsatellite loci in the conserved and fragmented sites

Maturational stage of *Magnolia obovata* plants	Site	Microsatellite loci											Average	Difference
		M6D1	M6D3	M6D4	M6D8	M6D10	M10D3	M10D6	M10D8	M15D5	M17D3	M17D5		
Adult	Conserved	19.26	12.61	7.66	4.61	4.56	12.31	6.26	9.16	2.96	7.99	7.17	8.60	
	Fragmented	11.55	9.96	9.80	4.46	5.08	10.35	5.98	7.56	3.24	8.03	5.11	7.37	p=0.0028
Sapling	Conserved	22.28	16.49	9.11	5.76	4.54	14.85	6.30	11.79	3.59	10.71	7.19	10.24	p=0.014
	Fragmented	14.81	7.09	8.88	4.50	5.06	11.95	5.53	7.60	2.92	8.58	4.84	7.43	

Table 2 Reproductive performance of *Magnolia obovata* in the conserved and fragmented sites

Traits	Site	Number of ovules/seeds			Ratio of (+)	χ^2	P
		+	−	Total			
Fertilization of ovules	Conserved	12,917	14,029	26,946	0.479	265.42	0.00
	Fragmented	16,649	13,761	30,410	0.547		
	Total	29,566	27,790	57,356	0.515		
Insect attack to seeds	Conserved	8,657	4,260	12,917	0.670	19.59	0.00
	Fragmented	11,560	5,089	16,649	0.694		
	Total	20,217	9,349	29,566	0.684		
Ovules that developed into seeds	Conserved	3,701	23,245	26,946	0.137	45.63	0.00
	Fragmented	4,787	25,623	30,410	0.157		
	Total	8,488	48,868	57,356	0.148		
Outcrossing at the stage of seeds	Conserved	103	92	195	0.528	2.95	0.09
	Fragmented	134	85	219	0.612		
	Total	237	177	414	0.542		

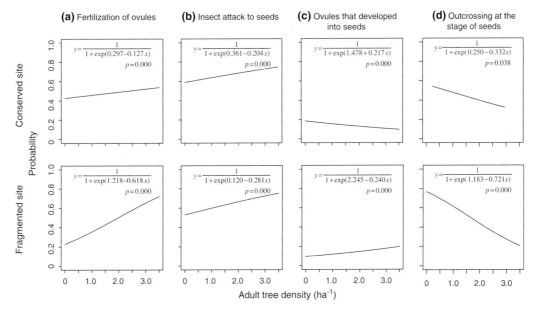

Fig. 2 Responses of fecundity traits to the density of adult *M. obovata* trees located within 200 m of a seed parent, at the conserved and fragmented sites

indicated by the randomized A & W test (Sork et al. 2005), there was no significant difference between the sites, suggesting equivalence in the amount of gene flow resulting from the movement of pollen. With an assumption of no local genetic structure among adult trees and no inbreeding between adults, the numbers of pollen donors (N_{ep}) were estimated with the equation $N_{ep} = 1/2\Phi_{FT}$ to be 3.6 for the conserved site and 3.5 for the fragmented sites.

Discussion

Genetic diversity

A decrease in the number of trees through deforestation will lead to a decrease in genetic diversity. The effective number of alleles is a more sensitive approach for detecting the immediate genetic loss following a population bottleneck than heterozygosity, which will be delayed until several generations after the disturbance. Although there were fewer effective alleles for the adult trees at the fragmented site than for adult trees at the conserved site, the difference was not significant ($P = 0.145$).

At the conserved site, the effective number of alleles for saplings was significantly ($P = 0.0028$) larger than that for adults. Until the conserved site was designated as a Forest Reserve in 1969, the forest had been subjected to partial cutting for charcoal-burning (Suzuki 2002), and the site presently consists of old growth and 50- to 80-year-old secondary forests. Therefore, the genetic diversity of *M. obovata* in this area likely decreased until 1969 as a consequence of the partial cutting policy, and the genetic diversity of the present adult trees in the conserved site must be lower, thereby reflecting the cutting in the past. Direct parentage analysis of saplings by means of microsatellite marker analysis revealed substantial gene flow into the conserved site, with 57% of genes of the sapling population derived from outside of the site (Isagi et al. 2000). Hence, the genetic diversity of the populationis probably recovering through active gene flow from outside of the site, and the larger effective number of alleles for saplings in the conserved site is likely to reflect the genetic recovery from degradation.

The effective number of alleles for saplings was significantly lower in the fragmented site than in the conserved site. This may reflect the larger gene flow into the conserved site than into the fragmented site that is currently underway, rather than a decrease in the number of effective alleles in the fragmented site itself. Since the TwoGener analysis did not reveal any significant differences in pollen movement between the conserved and fragmented sites, the observed difference in the gene flow seems to be caused by the difference in the number of seeds carried into the sites. The number of seeds carried into the fragmented site could be smaller than that carried into the conserved site because of the weaker attraction of the fragmented site for seed-dispersing birds.

Reproduction processes

In this research, we analyzed and compared processes of reproduction – from fertilization to germination – in *M. obovata* at both conserved and fragmented sites and found complicated responses of the reproductive traits to the processes to fragmentation. Although fertilization success was positively correlated to adult tree density at both the conserved and fragmented sites, the gradient of the curves for the fragmented site was much steeper (Fig. 2a). One possible explanation causing this pattern is that it would reflect a diminished efficiency of pollination at the fragmented site, with higher adult density compensating for the degradation of ecological function.

Because of severe inbreeding depression in *M. obovata* (Ishida et al. 2003), most of saplings were all out-crossed in the conserved site (Isagi et al. 2000), and the proportion of seeds that are out-crossed is important to population process of this species. The ratio of out-crossing at the seed stage was significantly correlated with adult tree density: the ratio of outcrossing was higher at a lower adult tree density. A similar pattern – a higher outcrossing ratio with lower density of plants or a longer distance of pollen movement with lower density – has been reported in earlier studies (e.g. Handel 1983; Fenster 1991; Godt and Hamrick 1993; Schnabel and Hamrick 1995; Kameyama et al. 2001).

Comprehensive evaluation of forest fragmentation

How can we evaluate the overall reproductive consequence of fragmentation based on the different responses of reproductive characteristics at the conserved and fragmented sites?

Fragmentation is thought to have negative ecological and genetic impacts on forest ecosystems. However, the effects of forest fragmentation may be subtle because of the long generation time of trees and the complex interactions of organisms in forest ecosystems. In the present study, we observed a rather complicated reaction to forest fragmentation in terms of genetic and fecundity traits such as effective number of alleles, ratio of fertilization of ovules, insect attack to seeds, ovules that developed into seeds, and outcrossing at the stage of seeds, among others. Because of the severe inbreeding depression in *M. obovata* (Ishida et al. 2003), outcrossed seeds predominantly contribute to the regeneration of the population. The ratio of ovules that developed into outcrossed seeds can be estimated from the product of the ratios of (1) ovules that developed into seeds and (2) outcrossing at the stage of seeds. This ratio showed a negative correlation with adult tree density: reproductive performance was higher with at a lower adult tree density at both the conserved and fragmented sites (Fig. 3). This correlation indicates the advantage of a low density of *M. obovata* trees in forest ecosystems and may imply a resilience to habitat fragmentation. Further comparative studies aimed at examining the correlation for other species growing at higher adult tree densities in natural conditions is required.

To date, most of the research that has been carried out on the effect of forest fragmentation has focused on reproductive traits such as fertilization, seed set and self-pollination rather than on regeneration success and regenerated population viability (Hobbs and Yates 2003; Tomimatsu 2005). The latest stage of the life history of *M. obovata* analyzed in this study was that of seedlings (TwoGener analysis). Although there were changes in reproductive traits related to fragmentation, the Two-Gener analysis did not detect any substantial changes in effective pollen movement and number of pollen donors

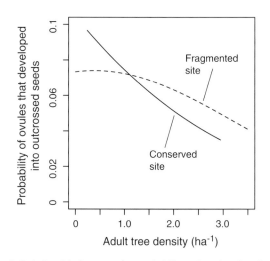

Fig. 3 Relationship between the probability of ovules that developed into outcrossed seeds and adult tree density within 200 m of a seed parent at the conserved and fragmented sites

for the seedlings. However, we have to advise caution with respect to the interpretation of the unchanged traits between the two sites based on the TwoGener analysis. The red seeds of *M. obovata* attract birds. In order to collect seeds before they were consumed by birds, it was necessary to climb and collect fruits high in the canopy. This constraint made it impossible to analyze large numbers of seed parents. There is a possibility that the relatively small number of seed parents used for the TwoGener analysis rendered the analysis incapable of detecting any difference (type II error) between the sites in terms of effective pollen movement, Φ_{FT}, and the number of pollen donors.

Do the present results indicate a resilience of *M. obovata* to the impact of fragmentation? In order to evaluate the effect of fragmentation on forest ecosystems comprehensively, we need to examine the impact of idiosyncratic factors – interactions between species and population processes – as well as issues relating to fecundity. Although any analysis of the consequences of fragmentation on seedling establishment and subsequent survival require longer study periods, they are directly connected to population persistence and are, therefore, important considerations when developing management guidelines for fragmented forests.

The pattern and amount of gene flow in plants are known to vary strikingly between species, years, populations, and even individuals (e.g. Ellstrand and Marshall 1985; Ellstrand et al. 1989; Streiff et al. 1999; Kameyama et al. 2000; Isagi et al. 2004). The flowering and fruiting of *M. obovata* fluctuates among years, and the activity of pollinators may also vary among years in response to weather conditions (Kikuzawa and Mizui 1990). Therefore, the effect of fragmentation needs to be assessed for several years. The size and spatial organization of remnants as well as the intensity and duration of the disturbance might also influence the impact of fragmentation. Additional research sites that are fragmented more severely than the ones studied here might yield more decisive results. Studies on species with a range of life history characteristics might also provide useful information by developing generalized management schemes for fragmented forest ecosystems.

One of the most important interactions to be considered when evaluating forest fragmentation is that between plants and pollinators. Although fragmentation has been regarded as a threat to pollination systems (Kearns et al. 1998; Duncan et al. 2004), unexpected patterns of pollen movement in degraded forests have been reported. Increased gene flow has been observed to occur after forest fragmentation, and pollinator behavior, wind dispersal, and accidental movement of small flying pollinators have been found to change following the opening of the canopy (Dick 2001; White et al. 2002). The complicated responses of fecundity found in the present study are likely to be influenced by the interaction between *M. obovata* and insects, with some of the latter acting as pollinators and others as seed predators. Changes in the composition of the species and the behavior of the pollinator and/or predator assemblages following forest fragmentation should be an important area of future research.

Acknowledgements The BIO-COSMOS Program of the Ministry of Agriculture, Forestry and Fisheries, Japan has supported research at the site area for 10 years. This research was partly supported by grants from the Ministry of Education, Science, Sports and Culture of Japan, the Research Institute for Humanity and Nature, and the 21st Century Center of Excellence Program at Hiroshima University.

References

Aizen MA, Feinsinger P (1994a) Forest fragmentation, pollination, and plant reproduction in a Chaco dry forest, Argentina. Ecology 75:330–351

Aizen MA, Feinsinger P (1994b) Habitat fragmentation, native insect pollinators, and feral honey bees in Argentine "Chaco Serrano". Ecol Appl 4:378–392

Aldrich PR, Hamrick JL (1998) Reproductive dominance of pasture trees in a fragmented tropical forest mosaic. Science 281:103–105

Aldrich PR, Hamrick JL, Chavarriaga P, Kochert G (1998) Microsatellite analysis of demographic genetic structure in fragmented populations of the tropical tree *Symphonia globulifera*. Mol Ecol 7:933–944

Austerlitz F, Smouse PE (2001) Two-generation analysis of pollen flow across a landscape. II. Relation between φ_{ft}, pollen dispersal, and inter-female distance. Genetics 157:851–857

Benitez-Malvido J (1998) Impact of forest fragmentation on seedling abundance in a tropical rain forest. Conserv Biol 12:380–389

Benitez-Malvido J, Martínez-Ramos M (2003) Impact of forest fragmentation on understorey plant species richness in Amazonia. Conserv Biol 17:389–400

Cascante A, Quesada M, Lobo JA, Fuchs EJ (2002) Effects of dry tropical forest fragmentation on the reproductive success and genetic structure of the tree *Samanea saman*. Conserv Biol 16:137–147

Collevatti RG, Grattapaglia D, Hay JD (2001) Population genetic structure of the endangered tropical tree species *Caryocar brasiliense*, based on variability at microsatellite loci. Mol Ecol 10:349–356

Cunningham SA (2000) Depressed pollination in habitat fragments causes low fruit set. Proc R Soc Lond B Biol 267:1149–1152

Dayanandan S, Dole J, Bawa K, Kesseli R (1999) Population structure delineated with microsatellite markers in fragmented populations of a tropical tree, *Carapa guianensis* (Meliaceae). Mol Ecol 8:1585–1592

Dick CW (2001) Genetic rescue of remnant tropical trees by an alien pollinator. Proc R Soc Lond B Biol 268:2391–2396

Dick CW, Etchelecu G, Austerlitz F (2003) Pollen dispersal of tropical trees (*Dinizia excelsa*: Fabaceae) by native insects and African honeybees in pristine and fragmented Amazonian rain forest. Mol Ecol 12:753–764

Duncan DH, Nicotra AB, Wood JT, Cunningham SA (2004) Plant isolation reduces outcross pollen receipt in a partially self-compatible herb. J Ecol 92:977–985

Ellstrand NC, Marshall DL (1985) Interpopulation gene flow by pollen in wild radish, *Raphanus sativus*. Am Nat 126:606–616

Ellstrand NC, Devlin B, Marshall DL (1989) Gene flow by pollen into small populations: data from experimental and natural stands of wild radish. Proc Nat Acad Sci USA 86:9044–9047

Fenster CB (1991) Gene flow in *Chamaecrista fasciculata* (Leguminosae) I. Gene dispersal. Evolution 45:398–409

Fuchs EJ, Lobo JA, Quesada M (2003) Effects of forest fragmentation and flowering phenology on the reproductive success and mating patterns of the tropical dry forest tree *Pachira quinata*. Conserv Biol 17:149–157

Godt MJW, Hamrick JL (1993) Patterns and levels of pollen mediated gene flow in *Lathyrus latifolius*. Evolution 47:98–110

Goudet J (1995) FSTAT version 1.2: a computer program to calculate F statistics. J Hered 86:485–486

Ghazoul J, Liston KA, Boyle TJB (1998) Disturbance-induced density-dependent seed set in *Shorea siamensis* (Dipterocarpaceae), a tropical forest tree. J Ecol 86:462–473

Handel SN (1983) Pollination ecology, plant population structure, and gene flow. In: Real L (ed) Pollination biology. Academic, New York, pp 163–211

Hedrick PW (2005) Genetics of populations, 3rd edn. Jones and Bartlett, Boston

Hobbs RJ, Yates CJ (2003) Impacts of ecosystem fragmentation on plant populations: generalising the idiosyncratic. Aust J Bot 51:471–488

Isagi Y, T Kanazashi W Suzuki H Tanaka T Abe (1999) Polymorphic DNA markers for *Magnolia obovata* Thunb. and their utility in related species. Mol Ecol 8:698–700

Isagi Y, Kanazashi T, Suzuki W, Tanaka H, Abe T (2000) Microsatellite analysis on regeneration process of *Magnolia obovata*. Heredity 84:143–151

Isagi Y, Kanazashi T, Suzuki W, Tanaka H, Abe T (2004) Highly variable pollination patterns in *Magnolia obovata* revealed by microsatellite paternity analysis. Int J Plant Sci 165:1047–1053

Ishida K, Yoshimaru H, Ito H (2003) Effects of geitonogamy on the seed set of *Magnolia obovata* Thunb. (Magnoliaceae). Int J Plant Sci 164:729–735

Judd WS, Campbell CS, Kellogg EA, Stevens PF, Donoghue MJ (2002) Plant systematics: a phylogenetic approach, 2nd edn. Sinauer Assoc, Sunderland

Kameyama Y, Isagi Y, Naito K, Nakagoshi N (2000) Microsatellite analysis of pollen flow in *Rhododendron metternichii* var. *hondoense*. Ecol Res 15:263–269

Kameyama Y, Isagi Y, Nakagoshi N (2001) Patterns and levels of gene flow in *Rhododendron metternichii* var. *hondoense* revealed by microsatellite analysis. Mol Ecol 10:205–216

Kearns CA, Inouye DW, Waser NM (1998) Endangered mutualisms: the conservation of plant-pollinator interactions. Annu Rev Ecol Syst 29:83–112

Kikuzawa K, Mizui N (1990) Flowering and fruiting phenology of *Magnolia hypoleuca*. Plant Species Biol 5:255–261

Lowe AJ, Boshier D, Ward M, Bacles CFE, Navarro C (2005) Genetic resource impacts of habitat loss and degradation; reconciling empirical evidence and predicted theory for neotropical trees. Heredity 95:255–273

Masaki T, Kominami Y, Nakashizuka T (1994) Spatial and seasonal patterns of seed dissemination of *Cornus controversa* in a temperate forest. Ecology 75:1903–1910

Moriguchi Y, Morishita T, Ohtani Y (2002) Climate in Ogawa Forest Reserve. In: Makashizuka T, Matsumoto Y (eds) Diversity and interaction in a temperate forest community. Springer, Tokyo, pp 11–18

Nakashizuka T, Iida S, Tanaka H, Shibata M, Abe S, Masaki T, Niiyama K (1992) Community dynamics of Ogawa Forest Reserve, a species-rich deciduous forest, central Japan. Vegetatio 103:105–112

Obayashi K, Tsumura Y, Ihara-Ujino T, Niiyama K, Tanouchi H, Suyama Y, Washitani I, Lee C, Lee SL, Muhammad N (2002) Genetic diversity and outcrossing rate between undisturbed and selectively logged forests of *Shorea curtisii* (Dipterocarpaceae) using microsatellite data analysis. Int J Plant Sci 163:151–158

Oostermeijer JGB, Luijten SH, Den Nijs JCM (2003) Integrating demographic and genetic approaches in plant conservation. Biol Conserv 113:389–398

Quesada M, Stoner KE, Rosas-Guerrero V, Palacios-Guevara C, Lobo JA (2003) Effects of habitat disruption on the activity of nectarivorous bats in a dry forest: implications for the reproductive success of the Neotropical tree *Ceiba grandiflora*. Oecologia 135:400–406

Quesada M, Stoner KE, Lobo JA, Herrerías-Diego Y, Palacios-Guevara C, Munguía-Rosas MA, O-Salazar KA, Rosas-Guerrero V (2004) Effects of forest fragmentation on pollinator activity and consequences for plant reproductive success and mating patterns in bat-pollinated bombacaceoos trees. Biotropica 36:131–138

R Development Core Team (2005) R: A language and environment for statistical computing. R Foundation for Statistical Computing, Vienna, Austria

Schnabel A, Hamrick JL (1995) Understanding the population genetic structure of *Gleditsia triacanthos* L. the scale and pattern of pollen gene flow. Evolution 49:921–931

Shibata M, Nakashizuka T (1995) Seed and seedling demography of four co-occurring *Carpinus* species in a temperate deciduous forest. Ecology 76:1099–1108

Slatkin M (1985) Gene flow in natural populations. Annu Rev of Ecol Syst 16:393–430

Smouse PE, Dyer RJ, Westfall RD, Sork VL (2001) Two-generation analysis of pollen flow across a landscape. I. Male gamete heterogeneity among females. Evolution 55:260–271

Sork VL, Smouse PE, Apsit VJ, Dyer RJ, Westfall RD (2005) A two-generation analysis of pollen pool genetic structure in flowering dogwood, *Cornis florida* (Cornaceae), in the Missouri Ozarks. Am J Bot 92:262–271

Streiff R, Ducousso A, Lexer C, Steinkellner H, Gloessl J, Kremer A (1999) Pollen dispersal inferred from paternity analysis in a mixed oak stand of *Quercus robur* L. and *Q. petraea* (Matt.) Liebl. Mol Ecol 8:831–841

Suzuki W (2002) Forest vegetation in and around Ogawa Forest Reserve in relation to human impact. In: Makashizuka T, Matsumoto Y (eds) Diversity and interaction in a temperate forest community. Springer, Tokyo, pp 25–41

Tanaka H (1995) Seed demography of three co-occurring *Acer* species in a Japanese temperate deciduous forest. J Veg Sci 6:887–896

Tanaka H, Yahara T (1988) The pollination of *Magnolia obovata*. In: Kawano S (eds) The world of plants, vol. 2. Kenkyusya, Tokyo, p 37

Thien LB (1974) Floral biology of *Magnolia*. Am J Bot 61:1037–1045

Tomimatsu H (2005) How do plant populations respond to habitat fragmentation (in Japanese)? Jpn J Conserv Ecol 10:163–171

White GM, Boshier DH, Powell W (1999) Genetic variation within a fragmented population of *Swietenia humilis* Zucc. Mol Ecol 8:1899–1909

White GM, Boshier DH, Powell W (2002) Increased pollen flow counteracts fragmentation in a tropical dry forest: an example from *Swietenia humilis* Zuccarini. Proc Nat Acad Sci USA 99:2038–2042

Wright S (1951) The genetical structure of populations. Ann Eugen 15:323–354

Young A, Boyle T, Brown AHD (1996) The population genetic consequences of habitat fragmentation for plants. Trends Ecol Evol 11:413–418

Ecol Res (2007) 22: 390–402
DOI 10.1007/s11284-007-0358-z

Naoki Agetsuma

Ecological function losses caused by monotonous land use induce crop raiding by wildlife on the island of Yakushima, southern Japan

Received: 24 February 2006 / Accepted: 11 December 2006 / Published online: 28 March 2007
© The Ecological Society of Japan 2007

Abstract Mass production is a logical outcome of price competition in a capitalist economy. It has resulted in the need for large-scale logging and planting of commercial crops. However, such monotonous land use, or monoculture, has damaged various ecological functions of forests and eroded the beneficial public service provided by forests. In Japan, the most widespread monotonous land use is associated with coniferous plantations, the expansion of which was encouraged by Forest Agency policies from 1958 that were aimed at increasing wood production. By 1986, half of all forested lands had been transformed into single-species conifer plantations. These policies may damage the ecological functions of forests: to provide stable habitats for forest wildlife. In particular, food supplies for wildlife have fluctuated greatly after several decades of logging. Some species have therefore changed their ecology and begun to explore novel environments proactively in order to adapt to such extreme fluctuations. Such species have started to use farmlands that neighbor the plantations. In this sense, crop raiding by wildlife can be regarded as a negative result of monotonous land use due to the loss of ecological functions. Therefore, habitat management to rehabilitate ecological functions and to reorganize the landscape will be required in order to resolve the problem of crop raiding by wildlife. This study examines crop raiding by Japanese deer (*Cervus nippon*) and monkeys (*Macaca fuscata*) on the island of Yakushima, which typifies crop-raiding situations in Japan.

Keywords Crop raiding · Functional response · Land use management · Plantation · Wildlife ecology

N. Agetsuma
Tomakomai Experimental Forest, Field Science Center
for Northern Biosphere, Hokkaido University, Takaoka,
Tomakomai 053-0035, Japan
E-mail: agetsuma@fsc.hokudai.ac.jp
Tel.: +81-144-332171
Fax: +81-144-332173

Monotonous forest and deterioration of ecological functions

Mass production is a logical outcome of price competition in a capitalist economy. It has necessitated intensive and monotonous land use for primary industries, especially large-scale logging, and the planting of commercial crops. Typical examples are the monocultures of coffee, palm, gum and sugarcane that are cultivated widely throughout tropical and subtropical regions (e.g., Nagata et al. 1994; Hartemink 2005). However, such monotonous land use, or monoculture, has damaged ecological functions and services provided by forest ecosystems, including those which benefit the public (e.g., McNeely et al. 1990; Lugo 1997).

In Japan, monotonous land use has developed mainly as the large-scale logging of natural forests and their replacement with coniferous trees, in accordance with the policies of the Forest Agency from 1958, which were aimed at increasing wood production (e.g., Japan Federation of Bar Associations 1991). Originally, Japanese forests consisted mainly of numerous broad-leaved species (i.e., summer green and evergreen broad-leaved forests: e.g., Miyawaki and Okuda 1978). However by 1986, half of all forested land had been logged and transformed into conifer plantations consisting of mainly single species, such as Japanese cedar (*Cryptomeria japonica*) and cypress (*Chamaecyparis obtusa*). Nonetheless, the financial situation of the Forest Agency administrating the National Forest worsened after 1973 for various reasons, including the start of mass importation of wood (e.g., Japan Federation of Bar Associations 1991). The situation for private forest owners was similar or worse. The cumulative debt of the Forest Agency in 1998 amounted to nearly ¥4,000 billion (ca. $30 billion). These policies have brought not only financial burdens to Japan, but they have also caused various environmental problems. In general, monotonous plantations exert various negative influences on the ecological functions of forest ecosystems. Conifer plantations in Japan decrease species diversity and disturb the species compositions of plant (e.g., Nagaike 2002) and animal communities (e.g., Saitoh and Nakatsu 1997; Maeto et al. 2002). Various material flows are influenced by such activities (e.g., Nakane 1995); mudslides occur frequently in logging and plantation areas (e.g., Inagaki 1999). Moreover, increasing numbers of Japanese people suffer allergies from cedar pollen (e.g., Inoue et al. 1986). These can be regarded as negative impacts of monotonous plantations through the deterioration of ecological functions and services.

Logging natural forests and transforming them into monotonous plantations profoundly disturbs the habitats of many mammals. For example, food supplies for herbivores tend to fluctuate greatly for around 20 years after logging and planting in Japan (e.g., Koizumi 1988; Sone et al. 1999; Hanya et al. 2005). Logging initially destroys food production, but it soon increases after improved exposure of the field layer to light. It then decreases rapidly to minimum levels, concomitant with tree growth (Fig. 1). These great fluctuations occur for two decades after logging. After this, productivity gradually recovers in a process that takes several decades or longer, even if vegetation regenerates naturally (Fig. 1a). Thus, forest mammals have been exposed to extreme fluctuations in food availability over long periods. Some species, such as deer (*Cervus nippon*), monkey (*Macaca fuscata*) and serow (*Capricornis crispus*), have begun to adapt various aspects of their ecology (diet selection, range use, daily rhythm, dispersion, reproduction, life history, social relation, etc.) and to proactively explore novel environments (farmlands neighboring the plantations and logging areas). By adapting to these new habitats, these mammals have started to damage agricultural crops (Agetsuma 1999b, 2006). In this sense, crop raiding by mammals can also be regarded as a negative impact of monotonous plantations.

A typical case is that of Yakushima Island in Japan, which was designated as a World Heritage site in 1993. As in most parts of Japan, large-scale logging and transformation to conifer plantations has engendered various environmental problems on this island. The diversity of woody plants and buried seeds has decreased in plantation forests (Aiba, unpublished data). Some insect taxa have also decreased in secondary forests and decreased greatly in plantations (Yumoto, unpublished data), while the diversity of foraging bat species has also decreased in plantations (Hill, unpublished data). Mudslides have occurred frequently in and around logging areas and young plantations (Japan Institute of Land and Environmental Studies 1981). These facts suggest that deterioration of ecological function due to the transformation to monotonous forest has occurred in Yakushima. Endemic subspecies of Japanese deer (*C. n. yakushimae*) and monkeys (*M. f. yakui*) inhabit the whole of this island. They have experienced great habitat disturbance as a result of the transformations. After extensive forest transformations (in the 1960s to 1970s),

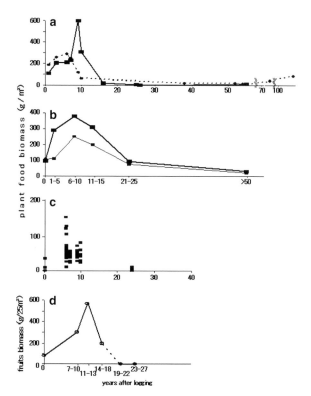

Fig. 1a–d Food plant biomass (dry weight/m²) after clear-cut logging of broad-leaved forests. **a** Food plant biomass under cedar plantations (*solid line*) and that under secondary broad-leaved forests for Japanese sika deer (*broken line*; modified from Takatsuki 1992). **b** Food plant biomass under conifer plantations in November (*solid line*) and in August for Japanese serow (*thin line*; drawn from Sone et al. 1999). **c** Food plant biomass under cedar plantations for Japanese serow (*squares*; modified from Ochiai 1996). **d** Biomass of fruit, which is an important food for Japanese monkey, in primary and secondary broad-leaved forests (*open circles*), and cedar plantations (*solid circles*; modified from Hanya et al. 2005)

these species began to intensively damage crops (in the 1980s to 1990s). This paper presents a review of the processes and factors associated with agricultural crop raiding by these mammals in relation to forest development in Yakushima, referring to cases from other locations in Japan. Then, the importance of land use management is examined to determine a possible route to recovering the damaged ecological functions, in order to prevent crop raiding as well as to conserve wildlife.

Forest development

Yakushima is a round mountainous island (ca. 500 km²) located in southern Japan (30°N, 130°E). About 14,000 residents populate its two townships. Most villages are located less than 100 m above sea level (a.s.l.); most other areas are forested (Fig. 2). About 80% of the forested area is National Forest property. From 0 to 800 m a.s.l., natural forests consist mainly of evergreen

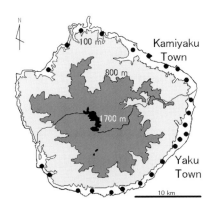

Fig. 2 The island of Yakushima. Villages are represented as *dots*. Contours show 100, 800 and 1,700 m a.s.l. A *line* shows the boundary between the two towns

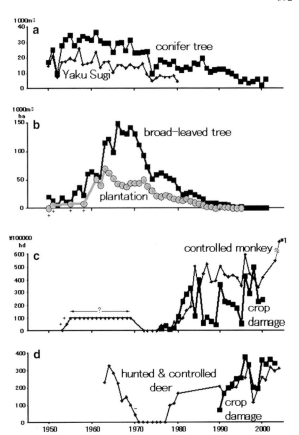

Fig. 3a–d Annual logging volumes, conifer planting area, agricultural crop damage by wildlife, and number of hunted animals. "+" indicates possibly underestimated values; "−" indicates possibly overestimated values; and "?" indicates approximate values in the figure. **a** Logging volume of conifer trees (*solid line*; Suwa, personal data) and those of cedar (*thin line*; Miura 1984; Fujimura 1971). These trees were logged at the higher altitudes of Yakushima. **b** Logging volume of broad-leaved trees (*solid line*; Suwa, personal data) and conifer plantation area (*gray line*; data from Kagoshima Pref.; data from Forest Agency). Broad-leaved trees were logged at both higher and middle altitudes. However, after 1963, they were mainly logged at middle altitudes. **c** Amount of crop damage (×¥100,000, ca. $800) by monkeys (*solid line*) and number of monkeys killed (*thin line*; Agetsuma 1998; Azuma 1984; Hirose 1984; data from Kagoshima Pref.). The value at *1 is 833 individuals. Monkeys were captured for experimental use in the 1950s and 1960s. After 1972, all captures were conducted as pest control measures. **d** Amount of crop damaged by deer (*solid line*) and the number of deer hunted and controlled (*thin line*; Kagoshimaken Shizen Aigo Kyokai 1981; Sueyoshi 1992; data from Kagoshima Pref.). During 1971–1977, deer hunting was prohibited. Since 1999, deer control has been concentrated around farms

broad-leaved trees. Then, from 800 to 1,800 m, forests consist of both broad-leaved and coniferous trees. There are many natural Japanese cedars, especially at over 1,200 m a.s.l. (Tagawa 1994). Cedars more than 1,000 years old are called "Yaku Sugi," and they produce very valuable timber (Kamiyaku Town 1984). Annual precipitation at lower altitudes is 2,500–5,000 mm, and that at higher altitudes is up to 7,000 mm, and occasionally 10,000 mm (Kagoshima Prefecture 1992). On the coasts, the annual mean temperature is around 20°C, which corresponds to the margin between subtropical and warm temperate zones (Tagawa 1994). On the other hand, above 1,000 m a.s.l., the climate is much cooler, with snow cover in winter.

Intensive logging in Yakushima began in the upper area for Yaku-Sugi. The logging of conifer trees increased around 1950 and decreased in the 1960s because of tree depletion (Fig. 3a). During 1955–1973, because the Japanese economy grew rapidly (average GNP growth rate was 9.1% per year according to data from the Cabinet Office of the Government of Japan), there was increased demand for wood and pulp production (Japan Federation of Bar Associations 1991). The logging area was then shifted to broad-leaved forests in the middle altitudes in Yakushima. The volumes of broad-leaved trees logged increased greatly from 1963, but decreased after 1973 (Fig. 3b), when the market price of wood dropped (Japan Federation of Bar Associations 1991), the number of trees that could be logged decreased, and the social trend of nature conservation became more popular (Kamiyaku Town 1984). After logging broad-leaved trees, Japanese cedar were widely planted in areas that were not the original habitat for the species (Fig. 3b).

Figure 4 shows logged areas in the National Forests of the island up to 1983. As the logging continued after 1983 (Fig. 3a,b), the actual total area logged must be much greater than shown. Most of the logging took place within a period of 20–30 years, and a quarter of all forested areas were transformed into monotonous conifer plantations (Kagoshima Prefecture 1992). The

plantations were particularly dense at middle-to-low altitudes on this island.

Land use changes in lower areas

Social and economic structures in regional societies in Japan, as well as those at a national level, are closely

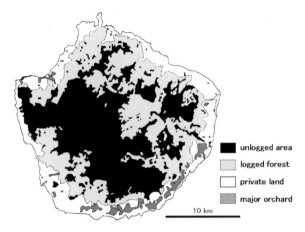

Fig. 4 Areas logged by 1983 and the main orchard in Yakushima (modified Agetsuma 1996). Very small unlogged forests among large logged forests were classified as logged areas. *Solid areas* indicate unlogged forests. *Lightly shaded areas* indicate logged areas. *Open areas* around coasts indicate private lands; *dark-shaded areas* indicate major orchards

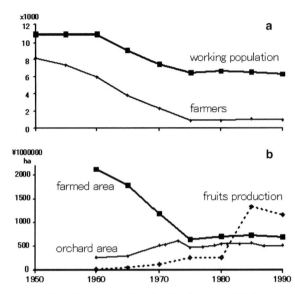

Fig. 5a–b Working population, farmland area and fruit production (×¥1000,000, ca. $8,000) in Yakushima. **a** Total working population (*solid line*) and number of people engaged in agriculture (*thin line*; Kagoshima Prefecture 1992). **b** Farm areas of the five main crop items (*solid line*); that of orchards (*thin line*); and fruit production (*broken line*; Agetsuma 1998; Kagoshima Prefecture 1992)

related to local land uses (Agetsuma 1999a). Below 300 m a.s.l. in Yakushima, the residents deforested the land and used it intensively for traditional purposes such as swidden cultivation, fuel wood, and charcoal (Sprague, unpublished data). Green fertilizer and litter for compost might also have been collected from the forest. As a result, "rough land," i.e., treeless land, spread extensively by the 1920s (Sprague, unpublished data). As both monkeys (Agetsuma 1995a, 1995b, 2001; Hanya et al. 2004) and deer (Agetsuma and Agetsuma-Yanagihara 2006) in natural forests in Yakushima depend largely on the production of tall trees, the treeless land might not be so valuable to them, although the deer could feed on herbaceous plants (Takatsuki 1990) and monkeys can feed on the fruits of some shrubs (Hanya et al. 2005). Higher human activity in farmlands would also have limited the utility of such places for animals. However, a fuel and fertilizer revolution then occurred. In addition, because of the shift in the industrial structure of this island, the number of farmers and the area of farmed land decreased greatly between 1950 and 1975 (Fig. 5a). In addition, some areas of treeless land and farm areas were rapidly converted into orange orchards around 1970 due to the promotional efforts of local governments (Fig. 5b; Agetsuma 1998). Some broad-leaved forests were also transformed into orange orchards at that time. Most treeless lands were abandoned and subsequently reverted to broad-leaved forests (Sprague, unpublished data), thereby providing many resources for animals after the 1970s. Consequently, the total area of treeless land decreased in the lower areas, while the areas of broad-leaved forests and orange orchards simultaneously increased. Production of oranges increased greatly after 1980 (Fig. 5b), but this industry has increasingly been damaged by wildlife.

Outbreaks of crop raiding and pest control

The population of raiding monkeys was estimated to be roughly 1,600–3,100 in 1991–1992 (see Table 1 of Yoshihiro et al. 1998, excluding Area 7 where there was no farmland), but there are estimated to be many other monkeys that bear no influence on crops in Yakushima (e.g., Hanya et al. 2004). The monkeys mainly damage oranges and other fruits, as well as sweet potatoes (data from Kamiyaku Town). Crop raiding by monkeys began before 1950 in Yakushima (Itani 1994). However, the amount of damage increased greatly after 1980, which was about 15 years after the peak of broad-leaved forest logging (Fig. 3b,c). Therefore, before the monkeys started to damage crops intensively, they had experienced great disturbances to their habitat through logging and planting.

Pest control of monkeys have been conducted from at least the 1910s (e.g., Kagoshima Dairinkusho 1916). After 1978, local governments greatly increased the number of monkeys controlled. However, from 1978 to 1983, even though more monkeys were controlled, there was still increased damage to crops (Fig. 3c). Thereafter, high levels of damage have continued to occur, even though 300–600 monkeys have been killed every year since 1984. It could be argued that an increase in the monkey population has been averted by controlling 300–600 monkeys since 1984. However, the number of monkeys removed has not been determined by scientific means, nor in response to the intensity of damage. Rather it has been mainly influenced by voluntary efforts of local hunters and the availability of subsidies from the local government. Therefore, it would be a lucky coincidence if the number of monkeys that were removed happened to be the number required to stop population growth. It seems much more likely that the rate of population increase varies to match the number of monkeys removed, thereby maintaining a fairly constant population size. The annual amount of damage by monkeys has varied greatly. Some of this variation might be attributable to interannual differences in the production of wild fruits (Noma, unpublished data), which are very important foods for monkeys during any season in the forests (e.g., Agetsuma 1995b, 2001; Hanya 2004).

Severe raiding of agricultural crops by deer in Yakushima has occurred since at least the eighteenth century (Kamiyaku Town 1984). In addition, Matsuda (1997) documented deer raiding fields in the 1950s. However, crop raiding by deer greatly increased around 1990 (Fig. 3d). This period was around 25 years after the logging peak of broad-leaved forests (Fig. 3b). The crops damaged were mainly orange trees (by bark stripping; data from the Kagoshima Prefecture), sweet potatoes and rice (Sueyoshi 1992). On the other hand, damage to planted conifer trees has decreased from the 1990s onwards (data from the Kagoshima Prefecture). This may be explained simply by the decline in newly planted areas after the 1970s (Fig. 3b) and the consequent decrease in the number of small planted saplings that are vulnerable to deer.

Deer hunting was traditionally conducted in Yakushima. By around 1950, more than 1,000 deer could be hunted per year (Kamiyaku Town 1984). However, the number of deer hunted decreased rapidly during 1964–1970 (Fig. 3d). Ultimately, deer hunting was banned from 1971, but was brought back for pest control from 1978. However, the damage caused by deer to crops and forestry did not seem to be substantial during this period (Tagawa 1987). Deer control was conducted over the whole island using very similar methods to those used when hunting before the ban, including at middle and even upper altitudes. Therefore, deer that took no interest in crops were also controlled. However, from 1999, deer control has been limited to areas around farmlands so that actual raiding deer are targeted. Since the number of deer controlled per year has stayed at around 200–300 since 1980 (Fig. 3d), raiding deer are now being controlled more intensively than before 1999. Nevertheless, the amount of crop damage has not decreased. As with the monkeys, raiding deer might also adapt to maintain their population size after depletion by intensive pest control.

Two common tendencies are recognizable in the crop raiding by both mammals in these empirical data. One is the timing of the rapid increase in crop raiding, which occurs after a delay of around 20 years from the logging peak of broad-leaved forests (Fig. 3b–d). Another is the ineffectiveness of pest control at reducing crop damage.

It has often been mentioned that pest control is ineffective at protecting crops from monkeys (e.g., Ministry of the Environment 2000; Hakusan Nature Conservation Center 1995). In addition, deer control appears to have had no beneficial effects in Yakushima to date.

Impacts of forest development on the ecology of mammals

The impacts of the transformation of natural forests into monotonous conifer plantations on the ecology of wildlife take a variety of forms, altering diet, habitat use and various behaviors (Gill et al. 1996). Some ecological functions are degraded by the transformation. The population density can be used as an easily detectable index to evaluate those impacts.

Monkey group density tends to have been degraded by the spread of plantations in Yakushima (Hill et al. 1994; Hanya et al. 2005). Food production of naturally regenerated young forests (< 20 years old approx.) for monkeys is higher than that of primary forests. However, that of plantation forests (> 20 years old) is very low in Yakushima (Hanya et al. 2005). Monkey group density corresponds closely to the food production of forests in Yakushima (Hanya et al. 2004; 2005). Various negative impacts of conifer plantations on Japanese monkeys have also been reported in other places. Monkey groups tend to avoid plantations within their ranges (e.g., Shirai 1993; Hakusan Nature Conservation Center 1995). Therefore, they must expand their ranging area after the establishment of conifer plantation nearby in order to compensate for that area's lower productivity (Furuichi et al. 1982). The topography of areas where cedar plantations have been established has made matters worse. Japanese cedar is typically planted in valleys and lower side slopes (e.g., Yamabayashi 1962), where food plant species diversity for monkey is high in Yakushima (Agetsuma and Noma 1995). Monkeys have therefore lost these diversified food resources, which they depend upon in spring when food conditions are poor in plantation areas.

Deer greatly depend on leaves and fruits of various woody species in various places at altitudes below 1500 m a.s.l. in Yakushima, although in some places they feed extensively on graminoides (Takatsuki 1990). In natural forest they are especially dependent upon fallen broad leaves (Agetsuma and Agetsuma-Yanagihara 2006) which arise mainly from tall trees in the canopy. However, a few tall broad-leaved trees might be able to regenerate in conifer plantations (e.g., Hanya et al. 2005). Therefore, the increase in food supply several years after logging (Fig. 1) might not be remarkable for monkey and deer in Yakushima, although they may be able to shift their diet to herbaceous plants in the logged areas.

Ohsawa et al. (1995) and Agetsuma et al. (unpublished data) surveyed deer density at four sites with

395

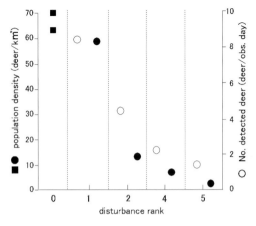

Fig. 6 Relative deer population density at six sites in Yakushima. *Solid squares* indicate estimated population densities in two lowland natural forests (50–200 m a.s.l.) almost free of plantations in August 2001 (Agetsuma et al. 2003). *Solid circles* indicate the estimated densities in the autumn of 2004 (Agetsuma et al., unpublished data) and *open circles* indicate the number of detected deer per observation day in the autumn of 1994 (Ohsawa et al. 1995) at four sites including plantations at middle altitudes (300–700 m a.s.l.). Sites with higher disturbance ranks have as more areas of younger plantations. Disturbance ranks 1–5 come from Hill et al. (1994)

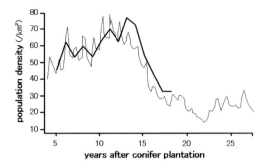

Fig. 7 Population change in roe deer (*Capreolus capreolus*) after the establishment of conifer plantations (rewritten from Gill et al. 1996). The *solid line* indicates the density estimated by the hypergeometric maximum likelihood method, and the *thin line* indicates the density based on the minimum number of known deer. Years after plantation are approximate values. Population density rapidly decreases around 15 years after intensive conifer plantation

different plantation occupancies in 1994 and 2004. In addition, Agetsuma et al. (2003) estimated the density in natural forests almost free of plantations in 2001. These studies showed that deer densities in heavily planted sites were much lower than those of relatively undisturbed sites (Fig. 6). This indicates that intensive plantations decrease deer density. The impacts cannot be cancelled out, even several decades after planting. Deer hunting and control statistics suggest that the deer population declined soon after intensive logging. Numbers of hunted deer decreased greatly in the 1960s (Fig. 3d). This decrease may indicate a decline in the deer population, but some local people thought that the decrease was caused by a decrease in hunting activity for social and economic reasons (Makise, personal communication). Kagoshimaken Shizen Aigo Kyokai (1981) also suspected that there was a notable decrease in the deer population in the late 1960s. On the other hand, the literature suggests that deer were commonly seen in forests around villages in 1925–1940 (Kamiyaku Town 1984; Miyamoto 1974). Therefore, it is reasonable to assume that deer populations fell precipitously in the 1960s and 1970s. Deer populations seemed to have recovered by the 1990s, as they began to be seen around villages once more (Fig. 7).

A similar decrease in the deer population caused by food depletion due to the establishment of conifer plantations has also been reported in other places (Gill et al. 1996). Roe deer (*Capreolus capreolus*) increased gradually over the 15 years following conifer planting, but decreased rapidly during the next few years; they then recovered very slowly (Fig. 6). This population dynamic corresponds to a great change in food availability after planting (Gill et al. 1996). A similar tendency is apparent in Japanese serow. Food availability decreased greatly 15 years after the establishment of conifer plantations (Fig. 1b). Changes in serow density corresponded to differences in food availability (Sone et al. 1999). However, these numerical responses to food availability show some time lags (Gill et al. 1996).

The recovery in the deer density might occur earlier at higher altitudes in Yakushima. Asahi et al. (1984) suspected that the deer density was higher at around 1,000 m a.s.l. than at lower altitudes after reviewing brief surveys of deer traces conducted in 1981. Intensive logging at higher altitudes began earlier than at middle altitudes (Kamiyaku Town 1984; Fig. 3a,b). In addition, at higher altitudes, much unlogged forest remains and conifer plantations are less widespread (Fig. 4). Therefore, food production might recover earlier in higher areas.

It is probable that populations of monkeys and deer decreased because of the lowered productivity caused by logging and planting. Subsequently, they gradually recovered as production recovered, mainly in naturally regenerating forest stands, from the 1980s. However, they might not recover to their original levels because the production from conifer plantations may still be low in the 2000s and, in fact, their population densities in plantation areas have also been much lower than in undisturbed forests (e.g., Hanya et al. 2005; Fig. 6).

Possible factors contributing to crop raiding

Several factors that may contribute to crop raiding and impacts on natural vegetation by wildlife have been suggested in Japan: extinction of natural predators; a decrease in the numbers of wild dogs; decreased snowfall; the expansion of the grasslands; and a decrease in the numbers of hunters (e.g., Matsuda et al. 1999; Miura 1999; Tsuji 1999; Agriculture, Forestry and Fisheries

Research Council et al. 2003; Hakusan Nature Conservation Center 1995). These factors have induced population eruptions of pest species. These provide the basis for adopting pest control as measure of this problem. Here, I examine the validity of these factors and their applicability to the case of Yakushima.

Extinction of wolves and decrease in numbers of wild dogs

The Japanese wolf (*Canis lupus*), a natural predator of herbivores, became extinct in 1905 in Japan. Generally in Japan, it is presumed that wolves regulated the number of deer and monkeys to "adequate" densities through predation. Consequently, the wolves must have served to limit the impact of these herbivores on crops and natural vegetation. Without their influence, herbivore populations would have erupted and completely destroyed vegetation (e.g., Stockton 2005; Peterson 1999). However, the effects of wolves on herbivore populations and their impact are not simple. Skogland (1991) reviewed the effects of predators on large herbivore populations using data from long-term studies. That study concluded that predators rarely regulate herbivore populations. Wolves may decrease deer when deer populations are small or decreasing, but they cannot limit large or increasing populations (e.g., Skogland 1991; Messier 1991). In fact, on Hokkaido Island (ca. 78,000 km^2) of Japan, more than 100,000 deer were hunted per year around 1870 when wolves still inhabited the island (Kaji 1995): therefore, the deer population must have been several times greater than 100,000, which would be far above the "eruption level" set by Hokkaido Prefecture (2002).

Some studies have reported that reintroduced wolves have decreased the feeding pressure from deer on natural vegetation (e.g., Ripple and Beschta 2004). However, it has also been pointed out that wolves may force deer to modify their range usage (Dussault et al. 2005), and consequently they relieve the intensity of the deer feeding pressure on vegetation in a particular location. In this sense, behavioral changes (i.e., functional response) of deer induced by wolves would be more important than the decrease in the numbers of deer through predation (i.e., numerical response). Moreover, some studies have shown that deer populations are regulated naturally without wolves and hunting, and that they do not unnaturally influence vegetation and nutritional cycles of soils very much, or even facilitate them (e.g., Singer et al. 1998; Boyce 1998).

Another large herbivore inhabiting Japan, the Japanese serow, has also lost its natural predator. In addition, in most places, the government passed laws prohibited serow hunting and control in 1934. Therefore, serow populations should have erupted since then without predation and hunting pressure. The population of this species did indeed increase until the 1990s. However, since then its population density has decreased

in various places (e.g., Nawa and Takayanagi 2001; Koganezawa 1999). There has been speculation that the decrease in the serow population is caused by severe competition with the increasing numbers of deer (e.g., Koganezawa 1999), but this seems unlikely, as decreasing tendencies are also apparent in areas without deer (e.g., Wildlife Workshop 2003; Miyazawa, unpublished data).

No large or middle-sized carnivores inhabited Yakushima originally (Environment Agency 1984). However, deer and monkeys have sustained their populations in forests which have a great diversity of plant species, including many endemic species (Tagawa 1994). Therefore, it is probable that some natural regulating mechanism functioned in Yakushima to sustain the forest ecosystem. For these reasons, wolf extinction does not seem to be a self-evident cause of the population eruptions of pest species in Japan. Moreover, the smaller body sizes of deer and monkeys in Yakushima are somewhat suggestive. Both deer and monkeys in Yakushima have smaller body sizes than other subspecies in Japan (Izawa et al. 1996). This feature may evolve under food limitation without predation (Kay 1998).

It has been anecdotally pointed out that a decrease in the number of wild dogs may have relaxed the limit on the deer population. However, if wild dogs perform similar functions to those of wolves, they are equally unlikely to regulate the deer population. The number of wild dogs seemed to increase from the late 1990s in the natural forest of Yakushima as the deer population recovered (Agetsuma, personal observation). In this sense, the wild dog population may be supported by the deer population, and their population dynamics might simply track those of deer.

Decreased snowfall due to global warming

Another factor might be a decrease in snowfall, attributable to global warming. Snowfall can be a limiting factor on herbivore populations (Skogland 1991). Heavy snows can result in a considerable decrease in population size. If snowfall decreases as a result of global warming, herbivore populations may increase sharply. However, we should note that global warming may not result in increases in the mean annual temperature; it may also increase the frequency of extreme weather (e.g., Sanchez et al. 2004), which might cause very severe winters. For example, many observatories in east Hokkaido have measured record snow depths since 2000 (data from the Japan Meteorological Agency). Especially in early 2004, 17 of the 36 observatories measured the heaviest snow depths in their recorded histories (which varied from 20 to 110 years depending on the observatory). Subsequently, deer populations seemed to decrease in 2004 (Fig. 8). However, the Hokkaido Prefectural authorities concluded that the data taken in 2004 must be underestimates (information from Hok-

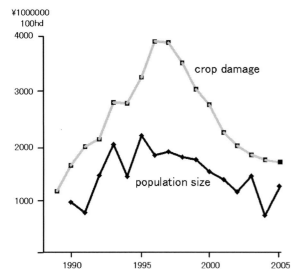

Fig. 8 Deer population and agricultural crop damage in the east part of Hokkaido Island, Japan. The *gray line* indicates the estimated deer population and the *solid line* indicates the amount of crop damage. The population size in 1994 was estimated at 200,000 by the Hokkaido Prefecture (2002). The populations in other years were calculated by averaging four relative density indices presented by the Hokkaido Pref.; sightings per unit effort from hunters (SPUE), density estimates from aerial surveys in main wintering places (three places for 1993 and 1996–2001, one place for 2002–2004), detection rates per 10 km of census routes, and those in main wintering places. Missing data were ignored. Since 1995, deer fences for protecting crops have been gradually elongating (over 3,000 km by 2003: Agriculture, Forestry and Fisheries Research Council et al. 2003; data from Hokkaido Pref.)

kaido Prefecture). Anyway, in 2005, the population recovered to or maintained their size at 2003 levels. It has been pointed out that expanding conifer plantations offer refuge to animals from snowfall (Kaji 1995). Therefore, the habitat structure might be a very important factor for deer populations, even in relation to the effects of snow. In the winter of 2005/2006, various places in Japan had their greatest ever recorded snowfalls (data from the Japan Meteorological Agency). Forthcoming data may confirm the impact of snow on wildlife populations.

The recent decreased levels of snowfall cannot explain the decline in the serow population. They tend to inhabit snow-filled areas and should be influenced by snow to a much greater extent than deer. In addition, heavy snows cannot be expected in warmer areas, including the middle to low altitudes of Yakushima. The number of days with snowfall per year in lowland Yakushima did not change much from 1939 to 2005 (0.36 day/year: 1939–1949; 0.70 day/year: 1950–1959; 2.33 day/year: 1960–1969; 0.56 day/year: 1970–1979; 0.60 day/year: 1980–1989; 0.20 day/year: 1990–1999; 1.00 day/year: 2000–2005), and the maximum snow depth was 2 cm (data from the Japan Meteorological Agency). The trend for the snowfall in the middle altitudes is likely to be similar.

Expansion of grasslands

The expansion of the artificial grasslands has been inferred to cause population eruption. Grasslands were often established after large-scale logging. Such grasslands produce more food for herbivores than do natural forests: they might therefore engender population eruption. However, the production is higher for 10–15 years at most; production then decreases rapidly to lower levels than those of natural forests for a much longer period (Fig. 1). Food conditions worsen thereafter, and population size may be limited by this lower production.

Forest roads were constructed during forest development. Some grassland strips occur at the roadsides. However, at the same time, plants cannot grow on the road surface itself or on concreted slopes around the road. Therefore, road construction has both positive and negative effects in terms of food production. Because roadside vegetation will also show successional change, the positive effects will decrease with time since construction. The forest roads also may not then have a great effect on the deer population, although they may exert some influence on the habitat use of individual animals.

The development of pastures yields much food for herbivores. However, for most parts of Japan, excluding Hokkaido Island, pasture areas are trivial (7% for Hokkaido, but 0.7% for other parts of Japan: data from The Ministry of Agriculture, Forestry and Fisheries) in comparison to the area of forested lands (71% for Hokkaido and 66% for other parts: data from the Forestry Agency). Also, in Yakushima, the area of pasture is less than 0.1% (Kagoshima Prefecture 2005), and that of the forest is 91% of the total area of the island (Kagoshima Prefecture 1992). For this reason, this factor may only be applicable to some limited areas that have many pastures. Even in such areas, habitat deterioration by logging and conifer plantations must be considered when gauging food conditions.

Some grasslands have been created unintentionally around farms. Such grasslands might attract wildlife and induce crop raiding. However, this influences habitat structures and range usage of animals rather than population eruption, even though increased population densities might occur in limited areas.

Decrease in the number of hunters

The number of registered hunters has decreased (Internal Affairs and Communications Agency 1992) concomitant with changes in social fashions in Japan. This is presumed to be a factor that explains the burgeoning populations of deer and monkeys (e.g., Tsuji 1999; Hakusan Nature Conservation Center 1995). About 30 professional hunters lived in Yakushima around 1950 (Kamiyaku Town 1984), but few professional hunters reside there now. Members of hunter associations in

Yakushima have also decreased (data from the Kagoshima Prefecture). However, we should note that hunting pressure cannot be assessed simply by the number of hunters. Hunting equipment and transportation (e.g., guns, transceivers, forest roads, automobiles, etc.) have improved considerably over the last 50 years. On the other hand, hunting skills and experience have likely declined, and hunting regulations have also undergone changes. Past hunting pressures on wildlife populations have not been adequately evaluated in Japan. Data reflecting the number of hunted animals exist in many areas in Japan, as shown in Yakushima (Fig. 3c,d). However, the task of interpreting these data is a complex one. Large numbers of hunted animals do not always imply strong hunting pressure on wildlife populations, because far more animals may remain alive, and vice versa. Notwithstanding, if the numbers of animals hunted were large over a long period, large animal populations would have to have been present to permit the steady supply of so many animals.

Deer hunting was strictly limited in Yakushima in the eighteenth century. Deer were controlled goods within the Shimazu Domain. People could not hunt deer freely (Kamiyaku Town 1984). In a particular village, deer hunts were conducted only three times each winter season. As a result, even in recent times, people eat venison only a few times per year (e.g., Kamiyaku Town 1984). Therefore, no proof exists that hunting pressure on deer or monkeys has ever regulated their populations in Yakushima. However, past hunting pressure should be examined more carefully by gathering as much local information as possible.

The need to seek other factors

Many Japanese researchers and administrators have inferred that the main cause of crop raiding is an "unnatural increase" in pest species because of the factors described above. Consequently, they may affirm that humans must control their numbers to "natural" or "adequate" levels to protect crops or even to conserve natural ecosystems. However, pest control measures have proved incapable of decreasing crop damage from various species of wildlife in various places. Direct control of wildlife has not mitigated their crop damage, implying that the factors that cause the problem have little to do with the mere number of animals involved. Nevertheless, so much emphasis has been placed on the mammal population eruptions that other factors and measures have not been seriously examined. It is therefore worth examining other factors and mechanisms of crop raiding from completely different points of view.

The putative factors mentioned above, excluding decreased hunters, could be adapted for limited areas or for limited wildlife species in Japan. However, agricultural crop raiding by various species increased across subtropical regions and cool temperate regions in Japan during the 1980s and 1990s (e.g., Agetsuma, 1999a). For

this reason, it is reasonable to seek factors that are generally applicable to crop raiding that can be adapted to various areas and to various species.

Relation between crop raiding and habitat transformations

Large-scale loggings of natural forest and transformations to monotonous conifer plantations occurred at similar times all over Japan (Japan Federation of Bar Associations 1991). Furthermore, the reconstruction of the structure of the landscape in forests and farm areas would have changed resource allocations for wildlife during the period when the Japanese industrial structure was changed. Such land use changes produce long perturbations that trigger changes in wildlife ecology. Therefore, this factor is applicable to all of Japan and to myriad pest species. In fact, a positive relation exists between the areal occupancy of conifer plantations and crop damage by deer in the Hyogo Prefecture (Sakata et al. 2001). In addition, crop damage by monkeys was high in places with 40–50% areal occupancy by conifer plantations (Japan Society for Preservation of Birds 1988). These imply negative influences of plantations on crop raiding.

In Yakushima, crop raiding increased rapidly for several years after 1980 for monkeys, and around 1990 for deer (Fig. 3c,d). However, because of the deterioration of food conditions after logging and conifer planting during those periods, their populations might have been lower than before intensive logging began. Therefore, wildlife population eruptions cannot explain the onset of intensive crop raiding. Some ecological and behavioral changes of animals are likely to cause crop raiding. If this was not the case, more crop raiding should have occurred previously (i.e., in the 1950s), when more animals inhabited the area and when much larger farm areas existed, than in the 1980s (Fig. 5). Moreover, it is impossible to explain the increased rates of crop damage at the beginning (mean 65% per year from 1976–1981 for monkeys, and mean 46% per year from 1990 to 1993 for deer; Fig. 3c,d) merely by population increases in pest species. Population growth rates of intensively provisioned monkeys are 10–14% (Oita City 1977). Those of deer without predation and hunting are reported to be 15–20% at most (e.g., Hokkaido Prefecture 2002), although a 36% increase was observed in one year after a population crash on Kinkazan Island (The Nature Conservation Society of Japan 1991).

Sometimes unclear or even reverse relationships between herbivore population densities and damage to crops, forestry and natural vegetation have been reported in Japan (e.g., Oi and Suzuki 2001; Ochiai 1996; Sakata et al. 2001), although the damage is commonly presumed to correlate with population densities (e.g., Suda and Koganezawa 2002; Hokkaido Prefecture 2002). In Yakushima, damage rates among wild plant species and plant individuals relative to deer density were much higher in plantation areas than in natural forests (e.g., Mupemo et al. 1999; Agetsuma et al. 2004). In east Hokkaido, damage caused to agricultural crops by deer increased rapidly during 1993–1998, even though the deer population was relatively constant (Fig. 8). Sakata et al. (2001) pointed out that some places with lower deer density receive greater crop damage and vice versa. Ochiai (1996) also suggested that damage of plantation trees by serow begins before their density increases. These empirical data and various other instances suggest that some functional responses (shifts in ecological strategies) play important roles in impacts on crops and vegetation, rather than simply the density of wildlife. We must note the importance of such functional responses and the ecological plasticity of wildlife if we are to understand and control their impacts. Attention must also be paid to the historical contexts of habitats and wildlife populations, not merely their current conditions (Gill et al. 1996).

The reasons why animals change ecological strategies and how they do so have never been seriously studied, but they can be explained as adaptations to fluctuations in habitat to some extent. Different ecological strategies have been understood as adaptations to habitat stability and unpredictability (e.g., Begon et al. 1986). Such adaptations explain interspecific differences in ecology. Similarly, the same species, and even the same individuals, within their respective ranges of capability, must change their ecologies in response to changes in habitat stability. Mammals have broad plasticity based on large behavioral repertoires with high abilities to learn and to physiologically adapt. These characteristics can allow them to shift their ecological strategies quickly upon great perturbations in habitat. In a fluctuating environment, generalist characteristics will be more adaptable than specialist characteristics. Therefore, mammals will tend to widen acceptable food species and ranges, and disperse to new areas in order to adapt to an unpredictable environment. As a result, they will tend to visit farmlands (novel places) and start to feed upon crops (novel foods). In addition, animals in less predictable environments will try to maximize their fitness within a short time frame because of the uncertainty over future benefits (e.g., Timberlake et al. 1987). Furthermore, optimum foraging theory (e.g., Charnov 1976) and experimental studies (e.g., Kacelink and Cuthill 1987; Agetsuma 1999c) show that animals facing lower food availability or dispersed food resources tend to feed longer at one food resource. These features engender intensive consumption of certain resources or resources in certain places, thereby causing intensive damage to crops and vegetation. These shifts in ecological strategies may be induced by large-scale transformations in habitat structure, which result in large perturbations.

Pest control must be an additional disturbance for wildlife populations. However, they seemed to have adapted to constant control pressures over the last 20–30 years in Yakushima because they have been able to continue damaging agricultural crops even under these pest control pressures (Fig. 3c,d). They may have retained their population size because similar control efforts since the 1980s have presumably resulted in a constant number of monkeys being controlled each year (Fig. 3c). If the size of the population of raiding monkeys has been constant (at 1,600–3,100 hd; Yoshihiro et al. 1998) since 1984, the population growth rate to compensate for the controlled animals (300–600 per year) would need to be very high compared to the intensively provisioned population growth rate (10–14% per year; Oita City 1977), even though that population size might be underestimated. Muroyama (2003) reported that monkeys that depend on agricultural crops year-round have birth rates as high as those of intensively provisioned monkeys. However, most raiding monkeys in Yakushima seem to feed on crops (oranges) for limited seasons.

Changes in landscape structure in forests and farm areas would facilitate crop raiding. In Yakushima's middle altitudes, highly productive broad-leaved forests have been transformed into low-productivity conifer plantations (Fig. 3b). On the other hand, at lower altitudes, low-productivity treeless lands have reverted to broad-leaved forest or have been transformed into farmland (Sprague, unpublished data). The altitudes associated with the production of resources for wildlife has changed inversely from middle-to-low altitudes to low-to-middle altitudes. Similar landscape changes are also apparent in other parts of Japan that are suffering from crop raiding (Agetsuma 2006). These changes in landscape structure encourage crop raiding and make it difficult to defend crops. Morino and Koike (2006) analyzed proximate environmental factors around orange orchards that influence crop raiding by monkeys in Yakushima. They showed that intensity of crop raiding was negatively correlated with the distance from large forest patches, and positively correlated with distance from wide roads. Therefore, appropriate landscape management could decrease crop raiding by wildlife, although it may not always promote ecological function around farmlands (Agetsuma 2006).

Habitat disturbances and changes in landscape structure that induce functional responses of wildlife will be important contributing factors to crop raiding. Even if population eruptions occur in some cases, distinguishing simple population recoveries from unnatural population eruptions must be very difficult; we note that it is accompanied by a shift in ecological strategies to large environmental perturbations. Takatsuki (1996) pointed out that pest control is only a tentative measure for crop raiding and that habitat rehabilitation is strongly required as an essential measure.

To confirm such shifts in ecological strategies and functional responses of mammals to habitat disturbances, we must analyze more histories of the disturbances wildlife have sustained in relation to their impacts on crops, forestry and vegetation. There must be many sources of empirical data throughout Japan, although they may not be perfect, because intensive forest development and crop raiding have occurred in many places. Moreover, research into the responses of individual animals to disturbances will provide much information on the shifts in their strategies.

Rehabilitation of ecological functions through land use management

Natural forests provide more stable environments with temporally and spatially diversified resources for various wildlife species than do artificially developed areas. These ecological functions help maintain the original ecologies of wildlife. Without them, wildlife will change diet selection, range use, dispersion, demography, etc., to have "pioneering" ecologies in order to adapt to large perturbations. This shift in ecological strategies may result in functional responses to disturbances. Species that cannot shift their strategies will falter numerically or become extinct. In fact, the strongest thread linking endangered species is habitat transformation (Hilton-Taylor 2000). In this sense, crop raiding and biodiversity loss express different aspects of the same ecological function losses. Therefore, habitat management (i.e., the rehabilitation of ecological functions and the reorganization of landscape structures) should have positive effects on both problems (Agetsuma 1995c, 1998, 2006). Nevertheless, few studies have been conducted from this point of view. The specific functions and processes needed to conserve wildlife ecology in forest ecosystems must be determined for taking actual measures against crop raiding.

Destruction of natural vegetation by deer has been reported in several parts of Japan, including Yakushima, as being a serious environmental problem (Yumoto and Matsuda 2006). To resolve this problem, we should note first that most natural vegetation has already been disturbed by forest transformation. Large-scale forest transformation deprives the original habitat of various plant species as a direct effect. These transformations may encourage deer to damage vegetation as a side effect. Therefore, habitat rehabilitation is required in order to conserve vegetation and natural ecosystems.

Presuming that there is an "ideal" or "expected" relationship among wildlife, vegetation and humans to conserve an ecosystem, we must devote careful attention to the relationships between them before performing heavy logging and planting. Most of our scientific knowledge of the dynamics of wildlife and forests has been obtained since the large transformations of the forests. Meanwhile, wildlife and plants have endured continual disturbances and have adapted to them. In more stable environments, or their original environments, they might show different dynamics and relationships. Attempts to conserve ecosystems and maintain ecological functions and services should make use of the many clues provided by the relationships among wildlife, natural vegetation and human activity

that occurred before the great forest transformations (Agetsuma 2005).

This study specifically addressed natural processes of agricultural crop raiding by wildlife in Japan. However, adaptations of society to this problem (through sociological, economic and political processes) also play an important role when elucidating the problem and prescribing its solution (Agetsuma 1999b).

Acknowledgments I thank Prof. Nakashizuka of Tohoku University for encouraging me to write this paper, the Forest and Fisheries Department of Kagoshima Pref. for providing me with local data, Mr. Suwa for providing data on forestry in Yakushima, Dr. Hill of the University of Sussex for improving the manuscript, and the PRI of Kyoto University for allowing me to use Yakushima Field Research Station. This study was supported by a grant from the Fujiwara Natural History Foundation, Grants-in Aid for Young Scientists (A) 14704013 and (B) 16780107 from The Ministry of Education, Culture, Sports, Science and Technology of Japan, and Research Projects of the RIHN.

References

Agetsuma N (1995a) Foraging strategies of Yakushima macaques (*Macaca fuscata yakui*). Int J Primatol 16:595–609

Agetsuma N (1995b) Dietary selection by Yakushima macaques (*Macaca fuscata yakui*): the influence of food availability and temperature. Int J Primatol 16:611–627

Agetsuma N (1995c) Methods of vegetation rehabilitation for wildlife conservation (in Japanese with English summary). Primate Res 11:133–146

Agetsuma N (1996) Nature conservation and wildlife in the island of Yakushima (in Japanese). Wildl Forum 2:23–32

Agetsuma N (1998) Crop damage by wild Japanese monkeys on Yakushima Island, Japan (in Japanese with English abstract). Jpn J Conserv Ecol 3:43–55

Agetsuma N (1999a) Present situation of Japanese wildlife reviewed from economic backgrounds: An introduction for young students. J Econ Dept Akita Univ Econ Law 30:11–23

Agetsuma N (1999b) Roles of primatology for wildlife management (in Japanese). In: Nishida T, Uehara S (eds) An introduction to field primatology. Sekaishisosha, Tokyo, pp 300–326

Agetsuma N (1999c) Simulation of patch use by monkeys in operant schedule. J Ethol 16:49–55

Agetsuma N (2001) Relation between age–sex classes and dietary selection of wild Japanese monkeys. Ecol Res 16:759–763

Agetsuma N (2005) Food web (in Japanese). In: Nakamura F, Koike T (eds) Forest sciences. Asakura, Tokyo, pp 80–83

Agetsuma N (2006) Conservation of wildlife (in Japanese). In: Field Science Center for Northern Biosphere, Hokkaido University (ed) Introduction to field sciences. Sankyo, Tokyo, pp 98–108

Agetsuma N, Agetsuma-Yanagihara Y (2006) Ecology of Yaku sika deer in the island of Yakushima (in Japanese). In: Ohsawa M, Tagawa H, Yamagiwa J (eds) World heritage, Yakushima. Asakura, Tokyo, pp 143–149

Agetsuma N, Noma N (1995) Rapid shifting of foraging pattern by Yakushima macaques (*Macaca fuscata yakui*) as a reaction to heavy fruiting of *Myrica rubra*. Int J Primatol 16:247–260

Agetsuma N, Sugiura H, Hill DA, Agetsuma-Yanagihara Y, Tanaka T (2003) Population density and group composition of Japanese sika deer (*Cervus nippon yakushimae*) in ever-green broad leaved forest of Yakushima, southern Japan. Ecol Res 18:475–483

Agetsuma N, Tsujino R, Kimura A, Kurotaki T, Baba K, Fukamachi N (2004) Evaluation of a forest as deer habitat (in Japanese). In: Sugiura H, Kaneko Y (eds) The 5th Yakushima field work course. Kamiyaku Town and Kyoto University, Inuyama, pp 14–18

Agriculture, Forestry and Fisheries Research Council, Forest and Forest Product Research Institute, National Agriculture and Bio-oriented Research Organization (2003) Basic knowledge for protecting products of agriculture and forestry from wildlife damages (in Japanese). Agriculture, Forestry and Fisheries Research Council, Tokyo

Asahi M, Izumi T, Nagai M, Hirabayashi T, Numaguchi K, Otsuka J (1984) Sika deer (*Cervus nippon yakushimae*) in the Yaku-shima Wilderness Area and its adjacent region, Yaku-shima Island, Kyushu, Japan (in Japanese with English summary). In: Environment Agency (ed) Nature of Yakushima. The Nature Conservation Society of Japan, Tokyo, pp 503–516

Azuma S (1984) Monkeys, forests and people (in Japanese). Monkey 28:94–102

Begon M, Harper JL, Townsend CR (1986) Ecology: individuals, populations and communities, 3rd edn. Blackwell Science, Oxford

Boyce MS (1998) Ecological-process management and ungulates: Yellowstone's conservation paradigm. Wildl Soc Bull 26:391–398

Charnov EL (1976) Optimal foraging, the marginal value theorem. Theor Pop Biol 9:129–136

Dussault C, Ouellet J, Courtois R, Huot J, Breton L, Jolicoeur H (2005) Linking moose habitat selection to limiting factors. Ecography 28:619–628

Environment Agency (1984) Nature of Yakushima (in Japanese). The Nature Conservation Society of Japan, Tokyo

Fujimura S (1971) Forest development and nature conservation (in Japanese). Water-Utility Research Institute Japan, Tokyo

Furuichi T, Takasaki H, Sprague DS (1982) Winter range utilization of a Japanese macaque troop in a snowy habitat. Folia Primatol 37:77–94

Gill RMA, Johnson AL, Francis A, Hiscocks K, Peace AJ (1996) Changes in roe deer (*Capreolus capreolus* L.) population density in response to forest habitat succession. For Ecol Manage 88:31–41

Hakusan Nature Conservation Center (1995) Researches on management of wildlife population and techniques to control their damages in agriculture, forestry and fisheries (in Japanese). Hakusan Nature Conservation Center, Yoshinodani

Hanya G (2004) Diet of a Japanese macaque troop in the coniferous forest of Yakushima. Int J Primatol 25:55–71

Hanya G, Yoshihiro S, Zamma K, Matsubara H, Ohtake M, Kubo R, Noma N, Agetsuma N, Takahata Y (2004) Environmental determinants of the altitudinal variations in relative group densities of Japanese macaques on Yakushima. Ecol Res 19:485–493

Hanya G, Zamma K, Hayaishi S, Yoshihiro S, Tsuriya Y, Sugaya S, Kanaoka MM, Hayakawa S, Takahata Y (2005) Comparisons of food availability and group density of Japanese macaques in primary, naturally regenerated, and plantation forests. Am J Primatol 66:245–262

Hartemink AE (2005) Plantation agriculture in the tropics. Environmental issues. Outlook Agr 34:11–21

Hill DA, Agetsuma N, Suzuki S (1994) Preliminary survey of group density of *Macaca fuscata yakui* in relation to logging history at seven sites in Yakushima Japan. Primate Res 10:85–93

Hilton-Taylor C (2000) 2000 IUCN Red List of threatened species. International Union for the Conservation of Nature, Gland, Switzerland

Hirose S (1984) Hunters of Yakushima (in Japanese). Monkey 28:82–87

Hokkaido Prefecture (2002) Management plan of Ezo sika deer (in Japanese). Hokkaido Prefecture, Sapporo

Inagaki H (1999) An occurrence of trees fallen by storm due to the difference of the vegetation and the following slope failures (in Japanese with English abstract). J Jpn Soc Eng Geol 40:196–206

Inoue S, Sakaguchi M, Mori H, Miyamura K, Ujiie A, Shigehara S, Noguchi Y (1986) Seroepidemiology of Sugi (Japanese cedar) pollinosis. Increase of IgE antibody positive rate in recent years (in Japanese). Igaku No Ayumi 138:285–286

Internal Affairs and Communications Agency (1992) Survey on status of wildlife conservation (in Japanese). Internal Affairs and Communications Agency, Tokyo

Itani J (1994) Nature and people in Yakushima (in Japanese). Seimei No Shima 30:25–29

Izawa K, Kasuya T, Kawamichi T (1996) The encyclopaedia of animals in Japan, vol. 2: Mammal II (in Japanese). Heibonsha, Tokyo

Japan Federation of Bar Associations (1991) Consideration of future forests (in Japanese). Yuhikaku, Tokyo

Japan Institute of Land and Environmental Studies (1981) Research on mudslides occurred in Kamiyaku Town, Kagoshima Prefecture (in Japanese). Kokudo Mondai 22:1–86

Japan Society for Preservation of Birds (1988) Survey on measures to control damages by wildlife: Japanese monkey and common cormorant (in Japanese). Environment Agency, Tokyo

Kacelink A, Cuthill IC (1987) Starlings and optimal foraging theory: modelling in a fractal world. In: Kamil AC, Krebs JR, Pulliam HR (eds) Foraging behavior. Plenum, New York, pp 303–333

Kagoshima Dairinkusho (1916) Statistics of Kagoshima Dairinkusho (in Japanese). Kagoshima Dairinkusho, Kagoshima

Kagoshimaken Shizen Aigo Kyokai (1981) Survey of distribution of Yaku sika deer (in Japanese). Kagoshimaken Shizen Aigo Kyokai Chousahoukoku 5:1–34

Kagoshima Prefecture (1992) Master plan of the Yakushima Environmental Culture Village (in Japanese). Kagoshima Prefecture, Kagoshima

Kagoshima Prefecture (2005) Annual statistics of Kagoshima Prefecture (in Japanese). Kagoshima Prefecture, Kagoshima

Kaji K (1995) Deer irruptions—a case study in Hokkaido, Japan (in Japanese). Mamm Sci 35:35–43

Kamiyaku Town (1984) Chronicle of Kamiyaku Town (in Japanese). Kamiyaku Town, Kamiyaku

Kay CE (1998) Are ecosystems structured from the top-down or bottom-up: a new look at an old debate. Wildl Soc Bull 26:484–498

Koganezawa M (1999) Changes in the population dynamics of Japanese serow and sika deer as a results of competitive interactions in the Ashio Mountains, central Japan. Biosphere Conserv 2:35–44

Koizumi T. (1988) Management of sika deer in Hokkaido. The effects of forest management and hunting on the deer populations (in Japanese with English summary). Res Bull Hokkaido Univ Forests 45:127–186

Lugo AE (1997) The apparent paradox of reestablishing species richness on degraded lands with tree monocultures. For Ecol Manage 99:9–19

Maeto K, Sato S, Miyata H (2002) Species diversity of longicorn beetles in humid warm-temperate forests: the impact of forest management practices on old-growth forest species in southwestern Japan. Biodiversity Conserv 11:1919–1937

Matsuda T (1997) Wonder stories on Yakushima (in Japanese). Shusakusha, Tokyo

Matsuda H, Kaji K, Uno H, Hirakawa H, Saitoh T (1999) A management policy for sika deer based on sex-specific hunting. Res Popul Ecol 41:139–149

McNeely JA, Miller KR, Reid WV, Mittermeier RA, Werner TB (1990) Conserving the world's biological diversity. IUCN, Gland; WRI, Baltimore; World Bank, Philadelphia

Messier F (1991) The significance of limiting and regulating factors on the demography of moose and white-tailed deer. J Anim Ecol 60:377–393

Ministry of the Environment (2000) Management manual for designated wildlife species (in Japanese). Ministry of the Environment, Tokyo

Miura K (1984) Nature conservation of Yakushima (in Japanese). Monkey 28:64–69

Miura S (1999) Wildlife ecology and damages on agriculture and forestry (in Japanese). National Forestry Extension Association in Japan, Tokyo

Miyamoto T (1974) Folklore of Yakushima (in Japanese). Miraisha, Tokyo

Miyawaki A, Okuda S (1978) Handbook of Japanese vegetation (in Japanese). Shibundo, Tokyo

Morino M, Koike F (2006) Assessing the risk of economic damage to crops caused by Japanese macaques on Yakushima Island (in Japanese with English abstract). Jpn J Conserv Ecol 11:43–52

Mupemo FC, Anantasran J, Harunari M, Hsu S, Kubo S, Ali W, Agetsuma N (1999) Population census of Yakushima deer. In: Yumoto T, Matsubara T (eds) Yakushima international field biology course. DIWPA, Kamiyaku Town, CER of Kyoto University, JCISE, Kyoto, pp 155–192

Muroyama Y (2003) To get along with monkeys around the village (in Japanese). Kyoto University Press, Kyoto

Nagaike T (2002) Differences in plant species diversity between conifer (*Larix kaempferi*) plantation and broad-leaved (*Quercus crispula*) secondary forests in central Japan. For Ecol Manage 168:111–123

Nagata S, Inoue M, Oka H (1994) Forest utilization pattern in the course of economic development. In an inquiry of the U-shaped hypothesis of forest resources (in Japanese with English summary). Rural Culture Association, Tokyo

Nakane K (1995) Soil carbon cycling in a Japanese cedar (*Cryptomeria japonica*) plantation. For Ecol Manage 72:185–197

Nawa A, Takayanagi A (2001) Population changes of Japanese serow and sika deer exposed by a point observation at southern part of the Suzuka Mountains, central Japan (in Japanese). Spec Publ Nagoya Soc Mammalog 3:58–63

Ochiai K (1996) Effects of forest management on Japanese serow populations in relation to habitat conservation for serows (in Japanese). Mamm Sci 36:79–87

Ohsawa H, Yamagiwa J, Hill DA, Agetsuma N, Suzuki S, Vadher SKA, Biggs AJ, Matsushima K, Kubo R (1995) Disturbances of ecology of wildlife by cedar plantation in Yakushima Island, WWF-J Report (in Japanese). Yakushima Wildlife Conservation Project, Inuyama

Oi T, Suzuki M (2001) Damage to sugi (*Cryptomeria japonica*) plantations by sika deer (*Cervus nippon*) in northern Honshu, Japan. Mamm Study 26:9–15

Oita City (1977) Survey on Japanese monkeys in Takasakiyama 1971–1976 (in Japanese). Oita City, Oita

Peterson RO (1999) Wolf-moose interaction on Isle Royale: the end of natural regulation? Ecol Appl 9:10–16

Ripple WJ, Beschta RL (2004) Wolves, elk, willows, and trophic cascades in the upper Gallatin Range of Southwestern Montana, USA. For Ecol Manage 200:161–181

Saitoh T, Nakatsu A (1997) The impact of forestry on the small rodent community of Hokkaido, Japan. Mamm Study 22:27–38

Sakata H, Hamasaki S, Kishimoto M, Mitsuhashi H, Mitsuhashi A, Yokoyama M, Mitani M (2001) The relationships between Sika deer density, hunting pressure and damage to agriculture in Hyogo Prefecture (in Japanese with English abstract). Hum Nat 12:63–72

Sanchez E, Gallardo C, Gaertner MA, Arribas A, Castro M (2004) Future climate extreme events in the Mediterranean simulated by a regional climate model: a first approach. Global Planet Change 44:163–180

Shirai K (1993) Home range and habitat use of Japanese monkeys in the plantation area in Okutama, Tokyo (in Japanese). Primate Res 9:300

Singer FJ, Swift DM, Coughenour MB, Varley JD (1998) Thunder on the Yellowstone revisited: an assessment of management of native ungulates by natural regulation, 1968–1993. Wildl Soc Bull 26:375–390

Skogland T (1991) What are the effects of predators on large ungulate populations? Oikos 61:401–411

Sone K, Okumura H, Abe M, Kitahara E (1999) Biomass of food plants and density of Japanese serow, *Capricornis crispus*. Mem Fac Agr Kagoshima Univ 35:7–16

Stockton SA, Allombert S, Gaston AJ, Martin J (2005) A natural experiment on the effects of high deer densities on the native flora of coastal temperate rain forests. Biol Conserv 126:118–128

Sueyoshi M (1992) Damages by sika deer on Yakushima, Kagoshima Prefecture (in Japanese). For Pests 41:33–35

Suda K, Koganezawa M (2002) Natural population density of sika deer considering forest ecosystem biodiversity (in Japanese with English abstract). Environ Res Quart 126:43–49

Tagawa H (1987) Researches on dynamics and management of biosphere reserves (in Japanese). In: Research on dynamics and management of Yakushima Biosphere Reserve. Research Team of "Dynamics and Management of Yakushima Biosphere Reserve", Kagoshima, pp 1–11

Tagawa H (1994) Natural World Heritage, Yakushima (in Japanese). Japan Broadcast Publishing, Tokyo

Takatsuki S (1990) Summer dietary composition of sika deer on Yakushima Island, southern Japan. Ecol Res 5:253–260

Takatsuki S (1992) Sika deer living in the North (in Japanese). Dobutsusha, Tokyo

Takatsuki S (1996) Conservation of common species (in Japanese). In: Higuchi H (ed) Conservation biology. University of Tokyo Press, Tokyo, pp 191–220

The Nature Conservation Society of Japan (1991) Wildlife conservation (in Japanese). The Nature Conservation Society of Japan, Tokyo

Timberlake W, Gawley DJ, Lucas GA (1987) Time horizons in rats foraging for food in temporally separated patches. J Exp Psychol 13:302–309

Tsuji M (1999) Can we protect forests of Nikko from deer damages? (in Japanese). Zuisousha, Utsunomiya

Wildlife Workshop (2003) Annual report of Serow population survey in Yamagata City (in Japanese). Wildlife Workshop, Yamagata

Yumoto Y, Matsuda H (2006) Deer eat World Heritage (in Japanese). Bun-ichi Sogo Shuppan, Tokyo

Yamabayashi N (1962) Forestry (in Japanese). Morikita, Tokyo

Yoshihiro S, Fruichi T, Manda M, Ohkubo N, Kinoshita M, Agetsuma N, Azuma S, Matsubara H, Sugiura H, Hill D, Kido E, Kubo R, Matsushima K, Nakajima K, Maruhashi T, Oi T, Sprague D, Tanaka T, Tsukahara T, Takahata Y (1998) The distribution of wild Yakushima macaque (*Macaca fuscata yakui*) troops around the coast of Yakushima Island, Japan. Primate Res 14:179–187

Ecol Res (2007) 22: 403–413
DOI 10.1007/s11284-007-0365-0

Sustainability and biodiversity of forest ecosystems:
an interdisciplinary approach

Masahiro Ichikawa

Degradation and loss of forest land and land-use changes in Sarawak, East Malaysia: a study of native land use by the Iban

Received: 25 February 2006 / Accepted: 11 December 2006 / Published online: 11 April 2007
© The Ecological Society of Japan 2007

Abstract Swidden agriculture, commercial logging and plantation development have been considered to be the primary common causes of degradation and loss of tropical rain forests in Southeast Asia. In this paper, I chose a part of northeastern Sarawak, East Malaysia as my case study area to analyze the changes in its land-use characteristics. In the study area, as well as primeval forests, we see that land use began about 100 years ago by a native group called the Iban; commercial logging began in the 1960s, and the development of oil palm plantations began recently. I describe the changes in land use as well as their social and economic causes by referring to aerial photographs, literature surveys, interviews with government officers and the Iban, and observation of land use. My analysis of land use demonstrates that on "state land", where commercial logging and oil palm plantation development are occurring, large areas of forest have been disturbed in a short period of time. The objective is to benefit economically in response to the social and economic conditions surrounding the study area. On the other hand, in the "Iban territory," where the Iban practice their land use, land conversion has not occurred on a large scale and in a short period of time, even though the forest has been cut and agricultural fields have been created in response to social and economic conditions as well. They disperse small agricultural fields throughout their forest land. Therefore, the landscape of the "Iban territory" is based on secondary forest, composed of patches of forest in various stages and with several types of agricultural land. Today in Sarawak, monocrop plantations are rapidly expanding and little primeval forest remains. Given these conditions, the land-use practices of natives such as the Iban will be evaluated from the viewpoint of ecosystem and biodiversity conservation. It could play

an important role in providing habitats for natural wildlife.

Keywords Tropical rain forest · Iban · Secondary forest · Swidden agriculture · Commercial logging · Oil palm plantation · Sarawak

Introduction

One of the primary causes of the degradation and loss of tropical rain forests, composed of complex ecosystems and the most diverse natures, is continuous and rapid conversion of primeval forest to other land uses (Walker and Steffen 1999). For the purpose of ecosystem and biodiversity conservation, it is necessary to examine the characteristics of these changes in land use. These could be, for example, the social conditions under which land-use changes occur and how the changes will evolve in the future. We should also consider the role of land use in areas other than primeval forests, because the amount of remaining primeval forest is already limited.

Tropical rain forests in Southeast Asia are mainly distributed in isolated areas of Malaysia, Indonesia and the Philippines. Although they are in different countries and there is distance between the areas, there are several commonalities with respect to forest degradation and loss. These include swidden agriculture, commercial logging (Lanly 1982), and plantation development (Primack and Corlett 2005), through which oil palm has become lucrative, particularly in the last two decades. Therefore, results of an analysis of land-use changes in a case study area can be applied to a wider area encompassing tropical rain forests distributed throughout Southeast Asia. This paper focuses upon Sarawak of East Malaysia on Borneo Island, which lies in the center of the distribution of tropical rain forests in Southeast Asia.

I give particular consideration to land use that includes swidden agriculture practiced by natives. Several researchers have considered native land use, especially

M. Ichikawa
Research Institute for Humanity and Nature (RINH),
457-4 Motoyama, Kamigamo, Kita-ku, Kyoto 603-8047, Japan
E-mail: ichikawa@chikyu.ac.jp
Tel.: +81-75-7072303
Fax: +81-75-7072507

swidden agriculture, as a cause of the degradation and loss of tropical forests (Lanly 1982; Freeman 1955). Based on those research results, after the 1960s, governments with jurisdiction over tropical rain forests have insisted that the main cause of the degradation and loss of tropical forests was swidden agriculture. On the other hand, they have stated that commercial logging, which they promoted, was a sustainable type of forest use because it has been conducted using selective logging.

Another researcher insists that if swidden agriculture is practiced in the traditional way with long enough fallow periods, it is a sustainable method of land use because the forest can recover (Chin 1987). Research results have found native land use including swidden agriculture to be positive (Aumeeruddy and Sansonnens 1994; Coomes et al. 2000; Salafsky 1993), because its mosaic pattern of fallow forests and various agricultural fields provides habitats for natural wildlife.

In Sarawak, oil palm plantation development began in the mid 1960s. From the 1990s in particular, the area used for plantations rapidly increased, while the production volume of timber decreased. There are many cases where oil palm plantations were developed after clear cutting forests where commercial logging had already been practiced. In recent years, plantation development using fast-growing trees, such as *Acacia mangium*, is planned or started in hilly or mountainous areas of the middle and upper reaches of river systems, which are inappropriate places for oil-palm growing. The rapid expansion of monocrop plantations in recent years has also been seen in Indonesia and the Philippines.

This paper addresses the issue of evaluating native land use from the perspective of its role in ecosystem and biodiversity conservation, at a time of change in the factors contributing to the degradation and loss of forests.

As already mentioned above, previous studies assessed native land use from the standpoint of the agricultural system, such as the use of fallow periods and periods of continuous cultivation, as well as from an ecological standpoint. In this paper, I focus on the scale and speed of the disturbance to forests, which have not been adequately considered, by analyzing periodical changes in land use. In conclusion, I highlight the important role of native land use for ecosystem and biodiversity conservation in the tropical rain forests of Southeast Asia, where the area used for monocrop plantations is rapidly increasing.

Overview of Sarawak, the selection of the study area, and methodology

Overview of Sarawak

Sarawak has a tropical rainforest climate, with an annual rainfall that ranges from 3,000 to 5,000 mm. The area of Sarawak State is about 120,000 km^2, of which

forest areas cover about 65% and fallow forests with patches of swidden cover about 30% (Department of Statistics, Malaysia 2004). The number of oil-palm plantations has been recently been on the rise, covering an area that occupies 4% of the state (Fig. 1). Untouched primeval forest is dispersed in small patches.

Before the Brooke (British Governor) regime began in Sarawak in the 19th century, primeval forest was mainly exploited by natives through swidden agriculture. Development in forests requiring larger amounts of capital proceeded gradually during the Brooke regime and during the British occupation after World War II. However, in the last few decades, development including commercial logging, oil-palm plantation development, road construction and urbanization, has proceeded rapidly.

Sarawak is home to about 2.7 million inhabitants (Department of Statistics, Malaysia 2004), and they are comprised of various ethnic groups, such as the Iban, Chinese, Malay and Bidayuh. The population is divided into two categories, natives and non-natives, such as Chinese and Indian. In Sarawak, almost everyone except the Chinese, who account for slightly less than 30% of Sarawak's population, are native peoples (Department of Statistics, Malaysia 2004). The largest ethnic group is the Iban who make up more than 30% of the population (Department of Statistics, Malaysia 2004).

Through Freeman's research, the Iban became well known as hill shifting cultivators (Freeman 1955). However, they usually practice wet rice cultivation in the flatlands (Ichikawa 2003a). More recently they have diversified their livelihoods to include the cultivation of cash crops and wage work (Ichikawa 2003b). In major parts of rural areas where other natives live, the residents have also practiced swidden agriculture, as the Iban have done, as one of their main sources of livelihood,

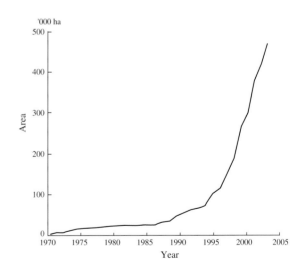

Fig. 1 Increase in the area of oil-palm plantations in Sarawak [source, Department of Agriculture Sarawak (1981, 1991); Department of Statistics, Malaysia, Sarawak (2004)]

although recently this is also changing. Therefore, it is generally true that with the exception of some groups, such as hunter-gatherers, land use by the natives in rural areas is similar.

Selection of the study area

The primary causes of the degradation and loss of tropical rain forests in Sarawak were noted earlier to be swidden agriculture, commercial logging and plantation development. Today, in rural areas in the middle and lower reaches of the rivers, we can see a mosaic landscape pattern consisting of undisturbed or less-disturbed forest and land use from the above-mentioned activities, i.e., fallow forests with swiddens, logged forests and oil-palm plantations. In this study, I chose an area of about 27,500 ha that encompasses the above-mentioned land uses (Fig. 2). It is located to the southwest, 25 km from Miri, an urban center northeast of Sarawak. The center of the study area is hilly or mountainous, and primeval forests still remain. The Bakong River, a tributary of the

Baram River, passes through the southeastern part of the study area. Iban land use extends along the left side of the river. On the right side, there are logged swamp forests, and some parts are beginning to be developed as oil-palm plantations. Oil-palm plantations are also seen in the southwest part of the study area and are beginning to be developed in the northwest as well. Logged forests extend through the northern part of the study area.

Methods

To analyze land use changes in my study area, I drew land use maps using several aerial photographs taken in 1963, 1977 and 1997, and measured the changes in land use. Classification of land use was conducted based on visual characteristics (Table 1).

Information on land use changes and their causes was collected by means of interviews and collection of documents. I conducted interviews with people familiar with the land use conditions both in the past and the present, such as the Iban and staff from the Forestry and

Fig. 2 The study area

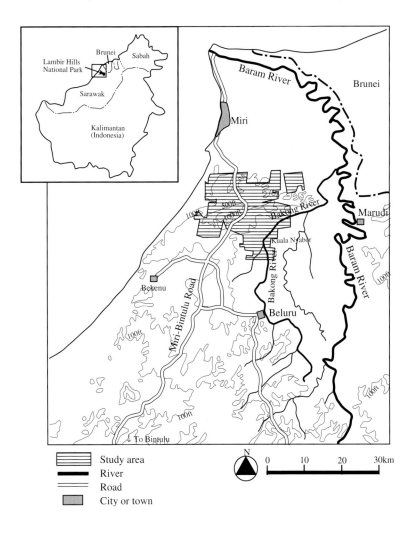

Study area
River
Road
City or town

Table 1 Characteristics of each type of land use visible on aerial photographs

Type of land use	Characteristics visible on aerial photographs
Primeval or mature forest in hills	Forest height is more than 30 m, and forest cover is almost 100%. No human disturbance has been made, or if there was some disturbance, the forest has recovered sufficiently
Peat swamp forest	Distributed in the lowlands along the Bakong River. Forest consists of *Shorea albida*
Swamp forest (primeval or mature)	Forest except for "peat swamp forest" extends through the lowlands along the Bakon River and in valleys among hills
Logged forest in hills	Selectively logged "primeval or mature forest on hills". Many logging roads visible. Tree density lower than in original forest
Logged forest in swamps	Selectively logged "peat swamp forest" and "swamp forests (primeval or mature)". Several railways for extracting logs can be seen. Tree density is lower than in original forest
Secondary forest in hills	Mainly fallow forest for swidden agriculture. Forest patches of variable height from 5 m to 30 m extend in a mosaic pattern. Swiddens, rubber gardens, fruit groves and pepper gardens 0.1–2 ha in size are dotted throughout this secondary forest
Secondary forest in swamps	Mainly fallow forest for swidden agriculture. Forest patches of variable height from 5 m to 30 m extend in a mosaic pattern. Swamp swiddens, and in relatively elevated places, rubber gardens and fruit groves 0.1–2 ha in size, are dotted throughout this secondary forest
Hill swidden	About 1 ha area with paddies, other crops, grasses and shrubs. Crops may be grown there, or were already harvested 1 or 2 years ago
Swamp swidden	About 1 ha area with rice, other crops, grasses and shrubs. Rice may be grown there, or was already harvested 1 or 2 years ago
Plantation	Area largely planted with para rubber and oil palm

Agriculture Departments, as well as Chinese living in towns near the study area. I visited the Iban five or six times each in ten longhouses located in the study area, and interviewed chiefs and seniors there. I collected documents, such as the Sarawak Gazette, earlier official gazettes, and statistics from government offices related to land use. The study was conducted intermittently from 1996 to 2005, and I lived in a longhouse for more than a year in total (cf. Ichikawa 2003c).

Results

Land use changes

The study area could be roughly divided into two types of area according to the landscape and the characteristics of the land-use changes (Fig. 3). In this paper, one is called "Iban territory" (around 7,600 ha), and the other is called "state land" (around 19,900 ha) (Fig. 3a).

The "Iban territory" consists of the areas where the Iban make their livelihood. These are almost all in "secondary forest in hills", "secondary forest in swamps", "hill swidden" and "swamp swidden,". The land code in Sarawak prescribes categories of land that the natives have the right to occupy and use. The category with the largest area is native customary right land, to which the natives have usufruct rights according to the adat of the natives, by cutting the primeval forest. The land code, however, limits the right to areas first cut before 1958. Only a few parts of the native customary right lands have been registered. Today, the natives sometimes use land without usufruct rights for their agriculture. The "Iban territory" in this paper consists of areas where the Iban consider themselves to have

usufruct rights and are actually still using the land. I confirmed those areas through interviews with the Iban. The Iban community lives together in a longhouse. The longhouse community territory primarily extends around the longhouse.

The "state land" consists of areas that have been used by the Sarawak state and private companies with permission from the government. In practice, some Iban have made swiddens on the "state land". Therefore, we actually see "secondary forest in hills" that contains fallow forest resulting from swiddens being fallowed, on the "state land" (Fig. 3c).

Land use changes in the "Iban territory"

There are two ways in which primeval forests were changed to secondary forests (Fig. 4). In one, the primeval forest was opened to make hill or swamp swiddens. The swiddens were fallowed after the harvest, and then secondary forest grew there. In the other, the primeval forest was selectively cut for commercial log production, and then swiddens were made there. After the swiddens were fallowed, they were replaced by secondary forest. From 1963 through to 1977, the area of primeval forest in both hills and swamps decreased, changing to secondary forest (Fig. 4a). In Iban territory, secondary forest eventually accounted for the largest part of the area (Fig. 4b).

Hill swiddens are made after cutting and burning a patch of forest on relatively dry land. The main crop is hill rice, and some vegetables, root crops and other crops are also produced (Ichikawa 2005a). In 1995, the cut forests for swiddens were from 7 to 10 years after the fallowing of the previous swiddens, and the forests were

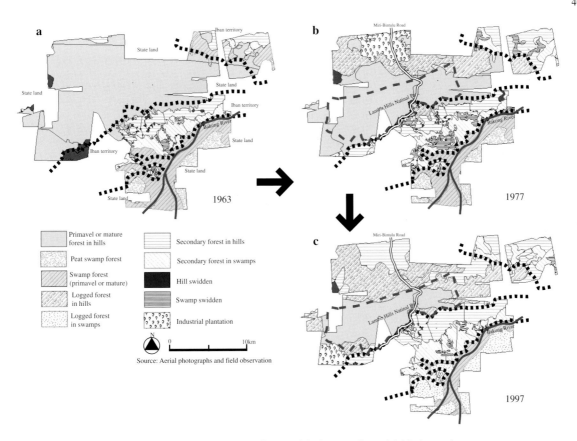

Fig. 3 Land-use changes in the study area, which are drawn from aerial photographs and field observation

about 10 m high (Ichikawa 2005a). Swiddens in swamps are made after cutting and burning forest or grass that grows on wet land (Ichikawa 2000). Only wet rice is produced there. Both types of agriculture incorporate fallow periods, although recently the period of time in which the same land is planted is becoming longer (Ichikawa 2000; 2005a). The area of hill swiddens decreased from 1963 to 1997 (Fig. 4b). In contrast, the area of swamp swiddens greatly increased from 1963 to 1977 (Fig. 4a), but then greatly decreased from 1977 to 1997 (Fig. 4b).

By 1997, the primeval forest and logged forest had almost all disappeared. Therefore, changes in land use today, as indicated by the arrows in the figure, primarily show changes between secondary forest and hill and swamp swiddens (Fig. 4b). As mentioned in Table 1, agricultural fields, such as rubber gardens, fruit tree groves and pepper gardens, are dispersed throughout the areas indicated as secondary forest. In 1995 a 2,600-ha area of the Iban territory containing agricultural fields was surveyed. According to interviews with individual owners, the total area of rubber gardens, fruit tree groves, and pepper gardens was estimated to be 220 ha, 38 ha and 3 ha respectively. Then, the total area of each type of agricultural field in the entire area of the "Iban territory" (7,600 ha) is roughly estimated to be 638 ha, 110 ha and 9 ha, respectively.

Interviews with the Iban revealed the land use changes as follows. Rubber gardens were created in the 1950s and in the early 1960s, although rubber planting started as early as the 1930s. Rubber gardens sometimes contained a few fruit trees in addition to rubber trees. There are also various naturally growing trees in the gardens, especially if tapping is not done for a while. According to Momose K et al. (unpublished data), 19 tree species/0.1 ha were counted in the rubber gardens in the study area. Pepper gardens were created in the early 1970s and in the late 1980s, although peppers were planted by a few families as early as the 1940s. A pepper garden is usually abandoned after about 10 years due to declines in production, and secondary forest grows on that land. The fruit tree groves were made by planting various kinds of fruit trees around work huts used primarily for rice preparation, but from the late 1980s onward, fruit trees were not necessarily planted only around the huts, but anywhere after secondary forest was opened. After 2000, several patches of oil palm gardens (1–2 ha in size) were created near the roads.

Causes of land-use changes in the "Iban territory"

The development of the Miri Division began after 1882 when Brooke annexed it as Sarawak territory (Pringle

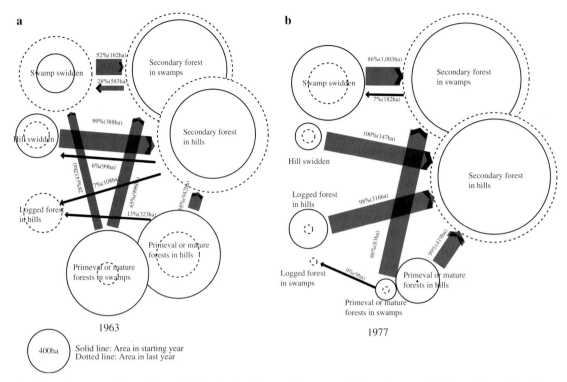

Fig. 4 Land-use changes and transit probability in the "Iban territory". Some legends on the land-use maps have been simplified to reduce complexity in the figures. The *circles* are proportional to the areas used for each type of land use. The *widths of arrows* are proportional to the percentage of change

1970). Many Iban longhouses are distributed along the upper to lower reaches of the Bakong River. The Iban came from today's Sri Aman and Sibu Divisions and settled there from the end of the 19th century to the early part of the 20th century (Pringle 1970; Sandin 1994). Before they settled there, much of the Bakong area was the territory of hunter-gatherers, and there was extensive forest that had never been cut. The original types of forest in the study area are peat swamp forest in the area south of the Bakong River, mixed swamp forest along the Bakong River, and mixed dipterocarp forest in the hills.

The Iban intentionally settled into the study area about 100 years ago, encouraged by Brooke (Pringle 1970). Due to land shortages, they wanted to move from what is today the Sri Aman Division to the Baram basin, where there was a lot of primeval forest. Brooke wished to stabilize the Baram basin, newly annexed territory, by immigrating the Iban. He also intended for the Iban to work as collectors of forest products, which were important trading commodities for Sarawak at that time.

The immigrants collected forest products, such as wild rubber and rattan, for sale, while cutting the primeval forest to make swiddens around their longhouses. Until the 1960s, hill swiddens were important for the Iban and larger than those in the swamps. This is because hill rice tasted and smelled much better than wet rice at that time, and because other products from hill swiddens were very important for subsistence under an economy more self-sufficient than that of today.

The reason for the increase in swamp swiddens from 1963 to 1977 (Fig. 4a) is an increase in rice demand, especially in the 1970s, from logging camps appearing in the Baram basin. Responding to the demand, the Iban expanded the swamp swiddens to produce commercial rice. In the 1990s, the field areas declined again (Fig. 4b) because products other than rice, such as pepper and forest products, became more attractive commodities, and there was an increase in opportunities to work for wages outside the villages, such as at construction sites in Miri. The reasons for the changes include the development of Miri, supported by prosperous oil and timber industries, as well as an increasing demand from urban dwellers for agricultural and forest products. The causes of the reduction in hill swiddens (Fig. 4b) include an increase in rice production from swamp swiddens after the 1970s, and a decrease in the importance of hill swidden products as it became possible to buy vegetables and root crops in Miri and other towns nearby.

Commercial logging in the "Iban territory" consisted of selective cutting on a relatively small scale conducted by small Chinese companies after the mid 1970s. Just after the logging, the Iban cut and burned the remaining trees to make swiddens. The companies continued to operate until the end of the 1980s, sometimes cutting the remaining thick trees in the "Iban territory".

Rubber and pepper have followed international market prices. They were planted when the prices were high, but in contrast, when prices were low, they were not adequately harvested and cared for. The Iban began to sell fruit in Miri after the 1980s, and in recent years fruit tree groves have been created. The reason why oil palm gardens were created was to be able to sell the products to nearby oil processing factories at a good price. The oil palm gardens were made by some Iban who had a certain amount of capital and land near roads where the products could be easily transported out. Therefore, the area of oil palm gardens made by the Iban is not expected to expand greatly in the near future.

Land-use changes in the "Iban territory" after 2000 appear to be of the following two types. In one, some patches of secondary forest are converted to swiddens, and then, after their fallowing, secondary forest regenerates. In the other, as with the example of the oil palm gardens, patches of new agricultural land are opened up, and forest products are collected in the areas shown as secondary forests, depending on social and economic conditions surrounding the study area.

Land-use changes on the "state land"

In 1963, almost all of the "state land" was covered by primeval forest (Fig. 5a). After the mid 1960s, companies conducted commercial logging in the hills. A rubber plantation was developed in the mid 1960s in a part of

the northern area of Fig. 3b. In 1975, the Lambir Hills National Park, which consists of about 7,000 ha, which had been a part of a forest reserve as a water catchment area, was established (Fig. 3b), and logging activities were banned in the area. As a result, primeval forests have been reserved there in good condition until today, although some illegal logging activities have been observed. In contrast, after the 1980s, companies have conducted commercial logging around the national park, not only in the mixed dipterocarp forest, but also in the peat swamp forest (Fig. 3c). The rubber plantation mentioned above was abandoned after some years of planting, and secondary forest regenerated there (Fig. 3c). Meanwhile, oil palm plantations have been developed in the southwestern part of the "state land" (Fig. 3c). Recently, oil palm plantations have been rapidly and widely developed in and around the study area (Fig. 6). After 2000, both logged forests in hills and swamps and secondary forest were clear-cut and converted into oil palm plantations.

Causes of land-use changes on the "state land"

In the study area, the development of forests on the "state land" began rather later than in the surrounding areas because of the steep topography. After the mid 1960s, the Miri-Bintulu road (MB road) acted as a trigger for commercial logging and plantation development (Fig. 3b). This construction project was promoted

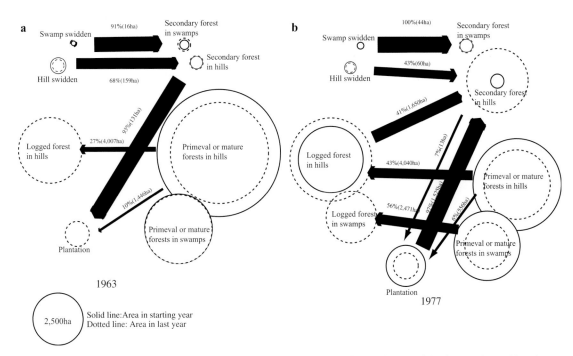

Fig. 5 Land-use changes and transit probability in the "state land". Some legends on the land-use maps have been simplified to reduce complexity in the figures. The *circles* are proportional to the areas used for each type of land use. The *widths of arrows* are proportional to the percentage of change

Fig. 6 Increasing monocrop
plantations in and around the
study area in 2002 (Source:
Land and Survey Department,
Sarawak)

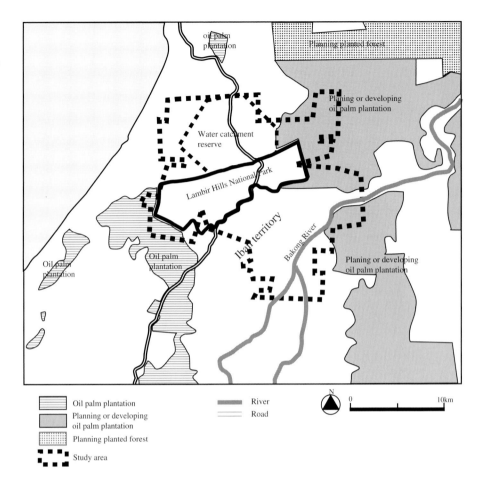

Oil palm plantation

Planning or developing
oil palm plantation

Planning planted forest

Study area

River

Road

N

0   10km

when Sarawak was incorporated into Malaysia in 1963. The road was constructed, and passed the study area in 1965. Soon after the road construction, many logging roads connecting to the MB road appeared in the northern part of the study area, and commercial logging by companies began in the hill dipterocarp forest (Fig. 3b). In fact, commercial logging in the hill areas increased after the 1960s throughout all of Sarawak (Ross 2001).

According to information from the Forestry Department, two logging licenses were issued to two different companies in 1966 and 1967 in the northern part of the study area, one for the western part of the MB road and the other for the eastern part of the road. Then the licenses were renewed several times. The one for the western part was still valid in 2005, and logging still continued. The other license for the eastern part was surrendered back in 1990, and there was no more logging after that. Every year before logging began, the logging companies submitted a "felling plan" to the Forestry Department, and they could cut trees only in the areas (coupes) shown in the plan. In a coupe, logging is not conducted just once, but rather repeatedly with several-year intervals. In the study area, the diameter of the logs gradually became smaller. I was not able to

obtain complete data showing how many times logging had been conducted in each coupe from 1966 or 1967. But some information shows that logging had been conducted two or three times with intervals of 6–8 years in the 1980s and 1990s in the area west of the MB road.

On the other hand, a global trend toward the conservation of nature appeared after the 1970s. According to staff at the Forestry Department, that was one of the reasons why national parks were designated in the areas where no logging and agricultural use had occurred.

Rubber plantation development was a governmental project intended to encourage immigrants from other regions of Sarawak to engage in rubber planting and management. However, the plantation project collapsed because the immigrants found wage-earning jobs, and the rubber plantations were abandoned and/or sold. According to the immigrants who still remain there today, after the mid 1960s, the price of rubber latex dropped, but several opportunities to work for wages arose, such as construction work in Miri and logging work in the Baram basin. After the collapse, secondary forest recovered almost everywhere, while in some areas fruit trees were planted.

Oil palm plantation development also began just outside of the study area when the MB road was constructed. Today, the largest oil palm plantation areas in Sarawak are found to the south of the study area. After the late 1990s, oil palm plantations began to be developed in the northern part of the study area (Fig. 6) after the logged forests were clear cut, although those do not yet appear in the aerial photographs from 1997 (Fig. 3c).

Discussion

Characteristics of land use on "state land" and in "Iban territory"

Land use changes in the "Iban territory" have generally proceeded more slowly than those on the "state land" (Fig. 7).

Until the mid 1960s, almost all of the "state land" was covered by primeval forest. After that, we see land use that generated great economic benefit in a short period of time. For example, one such use was commercial logging by companies. Many logging roads were constructed and thick wood was selectively cut using chainsaws. Logging was repeatedly performed in a coupe with short intervals of less than 10 years. In recent

years when logging resources declined, the forest was clear-cut in order to develop oil palm plantations. Developed lands were sometimes abandoned due to social and economic conditions, as seen in the case of the rubber plantation. Therefore, land use on the "state land" is characterized by changes in large areas of forest in a short period of time, as determined by the social and economic conditions of the time.

Economic development, however, has not been the only activity contributing to land-use change. National park gazetting is an example of a large primeval forest being preserved as a result of placing controls on logging activities by companies and agricultural activities by the Iban.

However, protected areas in Sarawak, such as national parks, comprise only 3,777 km^2 (3% of Sarawak state) (Department of Statistics, Malaysia 2004), and there is no trend toward significantly increasing the area in the future. In contrast, logged areas from 1963 to 1985 are estimated to be 28,200 km^2 (23% of Sarawak State) (Hong 1987), and after that period, the area increased. Oil palm plantations whose area is expanding rapidly today (Fig. 1) covered 4,648 km^2 (4% of Sarawak state) in 2003 (Department of Statistics, Malaysia 2004). Today, large plantations of monocrop trees, such

Fig. 7 Change in area for each type of land use: Iban territory (*upper*) and state land (*lower*). Areas in 1900 are estimated (Source, analysis of land use maps)

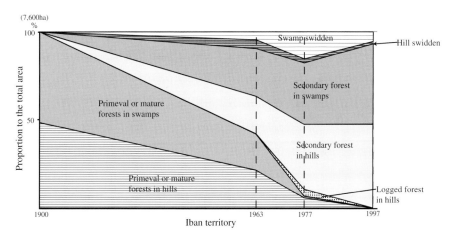

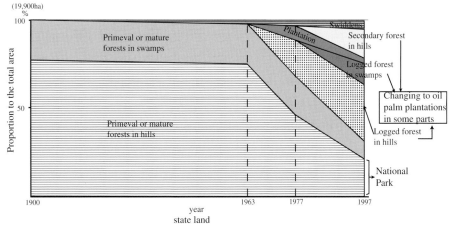

as acacia, have also been developed, or their development is planned for large areas of the middle and upper sections of the Sarawak river system. The state government aims at one million hectares of plantations within the next 15 or 20 years (Chan 1998).

On the other hand, in the "Iban territory", primeval forest was converted to secondary forest over a 100-year period (Fig. 7). However, the characteristics of the changes differ from those found on the "state land". Every year, each Iban household made swiddens consisting of a few hectares in a dispersed manner, opening up primeval forest and/or secondary forest. The swiddens were fallowed after cultivation for between one and several years. In recent years, since there is no remaining primeval forest, swiddens are created only in secondary forests.

The areas shown as secondary forests include small patches of reserved forests (*pulau galau*) which have never been clear cut (in order to ensure the availability of logs for house building), and small patches of various types of agricultural land, such as rubber gardens, fruit tree groves, oil palm gardens, and pepper gardens. Therefore, the land use in the "Iban territory" is based on secondary forests containing patches of vegetation at different stages, mixed in with patches of various kinds of agricultural fields (Ichikawa 2004). The patches form a mosaic pattern (Ichikawa 2004). As with the land use on the "state land", the land use in the "Iban territory" has been changing, influenced by the social and economic conditions around it. The land use there, however, has never resulted in large areas of land changing at once or in a short time, as seen on the "state land". These characteristics of Iban land use are sustainable over generations, and they are supported by the social institutions of the Iban, such as their system of land and natural resource tenure and of land inheritance (Ichikawa 2005b; Ichikawa unpublished data). We should note, however, that these institutions are also changing and are affected by social and economic conditions. The relationship between social and cultural factors and land use needs further study (see Ichikawa 2005b).

Evaluation of land use from the viewpoint of ecosystem and biodiversity conservation

In the rural areas of Sarawak, the best way to conserve the ecosystem and biodiversity may be to reserve primeval forest on a certain scale, because biodiversity of the primeval forest is outstandingly high (Lee et al. 2002). The land-use change with the most negative impact is conversion from primeval forest to large monocrop plantations. Information obtained by ecological research is insufficient to ascertain whether other land use, particularly the logged forests and secondary forests of the Iban, would be desirable for ecosystem conservation and biodiversity.

We may draw attention, however, to some points where the Iban land use is superior. With Iban land use,

relatively small-scale disturbances to the forest occur every year. Unlike the large-scale developments observed on the "state land", large areas of forest have not been converted in short periods of time. This land use, which is based on secondary forests, could be sustainable. Recent research in conservation ecology has indicated the significance of moderate disturbance for maintaining high biodiversity (Washitani and Yahara 1996). Other research revealed that land use by natives generally plays a much more important role in the conservation of ecosystems and biodiversity than do monocrop plantations (Primack and Corlett 2005). In the study area, 15–31 tree species/0.1 ha were observed in secondary forests, depending on the length of fallow period. In reserved forests, 47 tree species/0.1 ha were observed, while in primeval forest, 67 tree species/0.1 ha were observed (Momose K et al. unpublished data).

Today in Sarawak, the areas where commercial logging is finished have been rapidly converted to monocrop plantations, such as oil palm and acacia. In these conditions, the first priority for the ecosystem and biodiversity conservation will be to establish and manage protected areas, with careful attention to local communities around the area. However, the actual possibility of the establishment of protected areas will be quite limited. Considering these limited options for ecosystem and biodiversity conservation, land use by natives such as the Iban, who have often been blamed as vandals of primeval forests, should be reconsidered in a more positive light.

Acknowledgments This paper is the result of Research Project 2–2 at the Research Institute for Humanity and Nature (RIHN). I would like to thank Ms. Josephine Wong (Forestry Department, Sarawak), Dr. Mitsuo Yoshimura (Research Institute for Humanity and Nature), and Ms. Michi Kaga (former graduate student of Kyoto University) for their kind assistance in the acquisition of aerial photographs and the analysis of land-use changes. A portion of the fieldwork and the land-use mapping was funded by the above-mentioned project.

References

Aummeeruddy Y, Sansonnens B (1994) Shifting from simple to complex agroforestry systems: an example for buffer zone management from Kerinci. Agrofor Syst 28:113–141

Chan B (1998) Concerns of the industry on tree plantations in Sarawak. In: Chan B, Kho PCS, Lee HS (eds) Proceedings of Planted Forests in Sarawak, an International Conference, 16-17 February 1998, Kuching, Sarawak (vi–xii)

Chin SC (1987). Do shifting cultivators deforest? In: Forest resource crisis in the third world, Sahabat Alam Malaysia, Penang

Coomes OT, Grimard F, Burt G (2000) Tropical forests and shifting cultivation: secondary forest fallow dynamics among traditional farmers of the Peruvian Amazon. Ecol Econ 32:109–124

Department of Agriculture, Sarawak (1981) A digest of agricultural statistics, Kuching

Department of Agriculture, Sarawak (1991) Agricultural statistics of Sarawak 1990, Kuching

Department of Statistics, Malaysia, Sarawak (2004) Yearbook of statistics Sarawak 2002, Kuching

Freeman JD (1955) Iban agriculture: a report on the shifting cultivation of hill rice by the Iban of Sarawak. H.M.S.O, London

Hong E (1987) Native of Sarawak. Institut Masyrakat, Pulau Pinang

Ichikawa M (2000) Swamp rice cultivation in an Iban Village of Sarawak: planting methods as an adaptation strategy (in Japanese with English summary). Southeast Asian Studies 38(1):74–94

Ichikawa M (2003a) Shifting swamp rice cultivation with broadcasting seeding in Insular Southeast Asia. Southeast Asian Studies 41(2):239–261

Ichikawa M (2003b) Choice of livelihood activities by Iban household members in Sarawak, East Malaysia (in Japanese with English summary). Tropics 12(3):201–219

Ichikawa M (2003c) One hundred years of land use changes: political, social, and economic influences on an Iban village in the Bakong River basin, Sarawak, East Malaysia. In: Tuck Po L, De Jong W, Abe K (eds) The political ecology of tropical forests in Southeast Asia: historical perspectives. Kyoto University Press, Kyoto, pp 177–199

Ichikawa M (2004) Relationships among secondary forests and resource use and agriculture, as practiced by the Iban of Sarawak, East Malaysia. Tropics 13(4):269–286

Ichikawa M (2005a) Herbicide use in hill swidden agriculture and its background in Sarawak, East Malaysia (in Japanese). Tech Cult Agric (in press)

Ichikawa M (2005b) Inheritance of natural resources and their sustainable use by the Iban of Sarawak, East Malaysia—lands as a common resource among generations. Full paper submitted to the international symposium on Eco–human interactions in tropical forests organized by JASTE

Lanly JP (1982) Tropical forest resources. FAO Forestry Paper 30, FAO, Rome

Lee HS, Davies JV, LA Frankie JV, Tan S et al (2002) Floristic and structural diversity of mixed dipterocarp forest in Lambir Hills National Park, Sarawak, Malaysia. J Trop For Sci 14:379–400

Padoch C (1982) Migration and its alternatives among the Iban of Sarawak. KITLV, Leiden

Primack R, Corlett R (2005) Tropical rain forests. Blackwell, Oxford

Pringle R (1970) Rajahs and Rebels: the Iban of Sarawak under brooke rule, 1841–1941. Cornell University Press, Ithaca

Ross ML (2001) Timber booms and institutional breakdown in Southeast Asia. Cambridge University Press, Cambridge

Salafsky N (1993) Mammalian use of a buffer zone agroforestry system bordering Gunung Palung National Park, West Kalimantan, Indonesia. Conserv Biol 7(4):928–933

Sandin B (1994) Sources of Iban traditional history. The Sarawak Museum Journal 67, special monograph no 7

Walker B, Steffen J (1999) The nature of global changes. In: Walker B, Steffen J, Canadell J, Ingram J (eds) The terrestrial biosphere and global changes. Cambridge University Press, Cambridge, pp 1–18

Washitani I, Yahara T (1996) An introduction to conservation ecology (in Japanese). Bunichi sogo shuppan, Tokyo

Part 2
The basis and practice of sustainable management of forests and biodiversity

Ecol Res (2007) 22: 414–421
DOI 10.1007/s11284-007-0362-3

SPECIAL FEATURE

Sustainability and biodiversity of forest ecosystems:
an interdisciplinary approach

Peter Lagan · Sam Mannan · Hisashi Matsubayashi

Sustainable use of tropical forests by reduced-impact logging in Deramakot Forest Reserve, Sabah, Malaysia

Received: 21 December 2005 / Accepted: 24 October 2006 / Published online: 17 April 2007
© The Ecological Society of Japan 2007

Abstract In pursuance of economic growth and development, logging has exhausted the natural timber resource in the tropical rainforest of Sabah, Malaysia. Realizing the forest depletion, the Sabah Forestry Department, with technical support from the German Agency for Technical Cooperation, begun developing a management system with the intent of managing all commercial forest reserves in a way that mimics natural processes for sustainable production of low volume, high quality, and high priced timber products in 1989. As dictated by a forest management plan based on forest zoning, about 51,000 ha of the entire area is set aside for log production and 4,000 ha for conservation in Deramakot Forest Reserve, Sabah, Malaysia. This Forest Management Plan has served as the blueprint for operational work and biodiversity conservation in Deramakot to the present. A strict protection area is set aside for biodiversity conservation within the reserve. A reduced-impact logging system is being employed for harvesting with minimal impacts on the physical environment. Deramakot Forest Reserve was certified as "well managed" by an international certification body, the Forest Stewardship Council, in 1997 and is the first natural forest reserve in Southeast Asia managed in accordance with sustainable forestry principles. In addition to providing a "green premium," certification provides easier market access, evidence of legality, multi-stakeholder participation, conservation of biodiversity

and best forest management practices, particularly reduced-impact logging techniques. Deramakot Forest Reserve is the flagship of the Sabah Forestry Department and serves as a symbol of what can be achieved with political support and institutional commitment.

Keywords Deramakot Forest Reserve · Sustainable forest management · Reduced-impact logging · Tropical rain forests · Wildlife conservation

Introduction

Certification of forest management and labeling of forest products indicates that timber is legally produced from a sustainable source. Harvesting by reduced-impact logging method is used, giving careful consideration to vegetation loss and soil erosion. Major aspects considered during forest certification are the environment (conserving biodiversity and rare species, watershed protection, erosion control), the economy (costs and benefits), and society (involvement of local communities). The forest management is assessed by an accredited third-party external auditor every 6 months to ensure the continuance of certification and compliance with sustainable forestry practices. Some indicators, such as the level of biodiversity and the population of flagship species such as the orangutan, are required to monitor forest health. For example, a stable population of orangutans indicates that the management practices are nondeleterious. It is hoped that consumers will pay a premium for this added value of the forest products and to support sustainable forest management and conservation.

As part of the permanent commercial forest estate, the Deramakot Forest Reserve (05°15′28′N, 117°20′38′E) covers 55,083 ha of mixed dipterocarp forest in the east of central Sabah, Malaysia (Fig. 1). The climate is humid equatorial with a mean annual temperature of about 26°C and is greatly influenced by the northeast monsoon (November–February) and the

P. Lagan (✉) · S. Mannan
Sabah Forestry Department, Locked Bag 68,
90009 Sandakan, Sabah, Malaysia
E-mail: laganpeter@yahoo.co.uk
Tel.: +60-89-232600 + 60-89-232602
Fax: +60-89-232601

H. Matsubayashi
Center for Ecological Research, Kyoto University,
509-3 Hirano 2 chome, Otsu 520-2113, Japan

Present address: H. Matsubayashi
Tokyo University of Agriculture,
1737 Funako, Atsugi, Kanagawa 243-0034, Japan

Fig. 1 Location of Deramakot
Forest Reserve, Sabah,
Malaysia

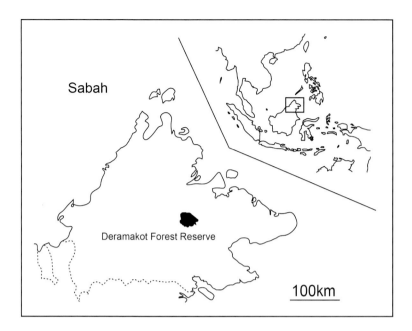

southwest monsoon (May–August). The average annual precipitation ranges from 1,700 to 5,100 mm (Kleine and Heuveldop 1993; Huth and Ditzer 2004).

The earliest known logging began in the southern part of the Deramakot, along the Kinabatangan River in the 1950s. The area was licensed for logging from 1955 to 1989. The minimum diameter trunk for harvesting was 60 cm and the felling cycle was 60 years. Loggers ignored the rule when it was convenient, attractive, and profitable. Variable cutting intensities of past management practices have resulted in an extremely heterogeneous condition of the remaining forest. Only 20% of the area is considered well stocked with harvestable trees, and more than 30% is covered by very poor forest with virtually no mature growing stock left.

Deramakot Forest Reserve was chosen in 1989 as the project site for the Malaysian–German Sustainable Forest Management Project for two reasons: it was the only logged natural forest that was neither licensed nor threatened by shifting cultivators, and the policy of the German Ministry of Economic Cooperation and Development prohibiting projects in pristine forests that involve timber harvesting. For the period 1989–2000, the Sabah Forestry Department, in collaboration with the German Technical Agency, implemented the Malaysian–German Sustainable Forest Management Project, which was made up of four phases: (1) a strong research emphasis with a component of management planning (1989–1992), (2) management planning, training and consolidation (1992–1994), (3) institution building, human resource and development, consolidation/implementation, and extension (1995–1998), and (4) consolidation, planning and human resource development (1999–2000). Deramakot Forest Reserve was certified as "well managed" by an international certification

body, the Forest Stewardship Council (FSC), in 1997. It is the first natural forest reserve in Southeast Asia managed in accordance with sustainable forestry principles.

A medium-term (10 years; from 1995 to 2004) forest management plan for Deramakot Forest Reserve was developed over a period of 5 years (1990–1994) as part of the project and was ready for implementation in 1995 (Forestry Department of Sabah 1995). We are now entering the second forest management plan for the next 10-year planning phase, encompassing the period from 2005 to 2014 (Forestry Department of Sabah 2005). The current and previous forest management plans have served as the blueprint for operational work in Deramakot Forest Reserve to the present.

Deramakot Forest Reserve is to be managed in accordance with sustainable forest management principles and a multiple-use approach to natural forest management. Amongst other things, the plan specifies that (1) not more than 20,000 m^3 are to be harvested each year (the annual allowable cut), (2) 1,000 ha are to be silviculturally treated each year, (3) 200 ha of rehabilitation planting per annum is to be carried out on degraded sites, (4) harvesting shall follow reduced-impact logging guidelines, (5) research and development will be conducted, and (6) eco-tourism shall be part of the plan implementation.

Deramakot Forest Reserve is divided into 135 compartments of varying sizes utilizing existing roads and other natural features (rivers, streams, ridges, foothills) as boundaries (Fig. 2). The forest entity is manned by 54 field personnel deployed over six major management activities: (1) harvesting (pre-harvest planning, harvest monitoring, and post-harvest planning or closing up), (2) road construction and maintenance, (3) silviculture,

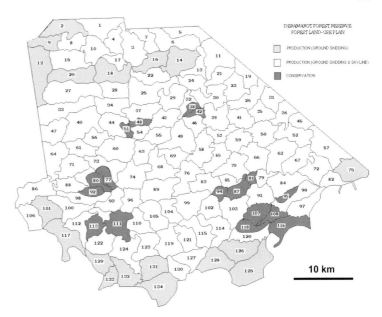

Fig. 2 Compartments of Deramakot Forest Reserve. *White* and *gray* indicate production areas. *Dark gray* indicates conservation areas

(4) rehabilitation, (5) administration, and (6) protection from illegal activity such as encroachment, hunting and forest fire.

In 1997, the Sabah Forestry Department engaged an auditing firm accredited by the Forest Stewardship Council to assess the management of Deramakot Forest Reserve under the Principles and Criteria for Forest Stewardship and the Malaysian Criteria and Indicators standard for sustainable forest management. The certification was successfully obtained covering a period of 5 years from July 1997 to July 2002. A major reassessment was carried out in Deramakot Forest Reserve by the same firm upon the expiration of the certificate in July of 2002. As a result, in April of 2003, Deramakot Forest Reserve was re-certified as a "well-managed forest" for a period of another 5 years from 2003 to 2008.

Forest management

The Deramakot Forest Reserve model owes its success to proper planning, concept development and to the implementation of the forest management plan. The objective is to manage the forest in a way that mimics natural processes for production of low volume, high quality and high priced timber. The main purpose of drawing up the Forest Management Plan is to define the 10-year planning objectives, which serve as guiding principles to plan ahead and operationalize the annual work plan. The main task of the Sabah Forestry Department is to prepare the annual work plan, which covers harvesting, silviculture, rehabilitation, and other forest management activities. The responsibility for supervising and monitoring all operations undertaken by the contractors lies with the Sabah Forestry

Department. Both the Sabah Forestry Department and the appointed contractors are jointly responsible for carrying out these operations and ensuring compliance.

Harvesting

Sustainability of timber harvesting means harvesting not more than the annual growth. Sustainable harvesting is a measure of economic viability and a criterion for ensuring self-sufficiency and profitable returns. The annual allowable cut of 20,000 m^3 was based on the individual tree growth simulation model, Dipterocarp Forest Growth Simulation Model (Ong and Kleine 1995). However, after 5 years in operation, the mid-term review conducted in 1999 recommended that the production volume be lowered to 15,000 m^3 to ensure sustainability. Another reason was that the annual allowable cut target was never met. Table 1 compares the annual allowable cut and actual volume harvested.

Extraction of harvested logs needs to minimize the mechanical impacts on the ground. We employ ground-skidding only on slopes less than 15° and skylines on slopes from 16° to 25°.

Silviculture and rehabilitation planting

Silviculture is essential because (1) the overall stocking of desirable commercial tree species is relatively low, (2) the infestation of climbing bamboo is high, and (3) it promotes growth and assists in natural vegetation.

From 1996–2001, a total land area of 1,146 ha was planted, which corresponds to 95.5% of the targeted 1,200 ha (200 ha per year). A crucial decision was made in late 2001 to stop the rehabilitation planting and in-

Table 1 Actual production versus the annual allowable cut

Year	Compartments	Annual allowable cut (m³)from the forest management plan	Actual volume harvested (m³)[a]
1995	73, 60	20,000	188.61
1996	73, 60, 49, 55	20,000	15,463.40
1997	73, 60, 49, 55, 68	20,000	13,794.16
1998	73, 43	20,000	12,235.95
1999	43, 63	20,000	914.80
2000[b]	43, 29, 44, 63	15,000	12,928.43
2001	44, 34, 37	15,000	10,741.83
2002	25, 37, 33	15,000	17,196.44
2003	12, 40	15,000	15,377.22
2004	40, 56	15,000	21,634.33
2005[c]	86, 47	17,600	7,721.00
Total		145,000	93,234.39

[a] Actual volume includes rejected logs, harvesting residue and logs used for bridge construction
[b] Mid-term review
[c] As of August 2005 (compartment 86 and part of compartment 47)

Table 2 Illegal felling and forest fires in the Deramakot area, 1995–2004

Year	Illegal felling (m³)	Forest fires (ha)
1995–1999	4,353	0
1997	–	250
2000	3,027	0
2001	214	0
2002	15	0
2003	0	0
2004	0	0
Total	7,609	250

stead to concentrate on maintaining planted seedlings for financial reasons and because of a mismatch between site and species. Those seedlings that were properly planted on suitable sites demonstrate low mortality and greater growth rates.

Protection

Protection is emphasized to ensure the security and stability of Deramakot Forest Reserve. The boundaries of Deramakot Forest Reserve, particularly those bordering alienated lands, were demarcated in 2002. Properly demarcated boundaries will facilitate enforcement work. Illegal felling has occurred over the years with the most serious occasions involving tractors. By and large this has subsided (Table 2), and if it occurs, will most probably be confined to small-time riverine felling, a form of cultural harvesting unique to the riverine communities along the Kinabatangan River.

The other threat to the forest is forest fires. The forest fires that originate from human ignitions are difficult to suppress at the best of times. In Deramakot Forest Reserve, a fire management plan is operationalized during fire season. We have an early detection and warning system (weather monitoring, fire index), prevention measures (education and awareness), and fire suppression equipment in place.

Research and development

Many scientific papers covering various fields (ecology, entomology, mammalogy, hydrology, silviculture, harvesting) have been written based on research conducted in Deramakot Forest Reserve, and many more are expected to be published in the future. Under the Eight Malaysia Plan, research on harvesting is being conducted, whereby various parameters (diameter limits, slope limitations, comprehensive harvest plan preparation, etc.) will be investigated. Deramakot Forest Reserve attracts a fair number of local students each year who conduct practical coursework prerequisites in the reserve. International researchers also visit Deramakot. Noteworthy is a team of Japanese researchers who investigated the impact of logging on biodiversity of tropical rain forests under a project of the Research Institute for Humanity and Nature. Japanese researchers in collaboration with the Sabah Forestry Department have demonstrated the efficiency of reduced-impact logging in sustaining the diversity of several taxonomic groups at the level of a pristine forest (Lee et al. 2006).

Social responsibilities

For sustainable forest management, the Forest Stewardship Council's principles and criteria address the participation of the local and indigenous people living within or on the fringes of forest reserves.

There are no indigenous people living inside Deramakot Forest Reserve, but there are six villages (20–50 households each) located on the southern fringe of Deramakot Forest Reserve along the Kinabatangan River, the longest river in Sabah. Their livelihood involves freshwater fishing; cultivating dry-paddy rice, cassava, and maize; and collecting non-wood forest products, such as rattan and medicinal plants.

A committee was set up specifically to address problems and issues of the local communities in relation to forest management in Deramakot Forest Reserve.

Meetings with representatives of the villages were conducted three times a year. Discussions primarily focused on improving the villagers' well-being so that they are able to earn a living without having to abuse the forest. Job employment as forest workers for the various forest management activities (harvesting, planting, silviculture, and boundary demarcation) in Deramakot Forest Reserve and home stay for eco-tourists were considered. Human–wildlife conflicts involving the destruction of cultivated land by elephants and other wildlife were also discussed. The welfare of the local communities, alleviating them from poverty and hardship, is the main issue, starting with supplying gravity-fed potable water (where the source is from Deramakot Forest Reserve) to one village and the creation of jobs.

Natural forest management under forest stewardship gives the greatest promise for rural jobs attuned to the cultural norms of the forest inhabitants. Silviculture, an important tool of natural forest management, suits our indigenous people and rural natives. In Deramakot Forest Reserve, 100% of silvicultural workers are natives, who earn a monthly wage. The natives from these villages are also employed in demarcating the boundary along sensitive areas of the forest.

Wildlife management

Mitigating the impacts of forest management activities on wildlife

As planned in the Forest Management Plan, approximately three-quarters of Deramakot Forest Reserve remains undisturbed or closed to forest management activities at any given time. This means all forest management activities (silviculture, enrichment planting, and harvesting) are focused on a small portion (10,000 ha) of the Deramakot Forest Reserve staggered over a period of 10 years, which translates to a management cycle of about 40–50 years. This is planned primarily to encourage plant succession without disturbance, and at the same time, the undisturbed areas act as a sanctuary for wildlife that thrives in Deramakot Forest Reserve.

Wildlife and their habitat contiguity are ensured in Deramakot Forest Reserve to ensure their sustainability. In addition, mitigating measures (Table 3) minimize the impact of human presence and interference with the ecosystem.

Wildlife conservation and monitoring

An integral part of the forest is its fauna resources. Wildlife in Deramakot Forest Reserve has received little attention in the past as the primary objective was timber management. Timber production will remain the dominant factor in planning land use in Deramakot Forest Reserve, however, to meet the requirements under principle #9 [identification of high conservation value forest (HCVF)] of the Forest Stewardship Council's principles and criteria, wildlife is increasingly gaining importance in sustainable forestry.

Two major studies on wildlife have been conducted in Deramakot. One is a population estimate of orangutans by aerial survey in 1999 (Ancrenaz et al. 2005). The other is the survey of mammal fauna at natural licks by camera trapping from 2003 until 2005 (Matsubayashi

Table 3 Mitigating the impacts of forest management activities on wildlife in Deramakot Forest Reserve

Activity	Impacts	Current management practice (mitigation)
Road construction and maintenance	Soil erosion. River/stream sedimentation. Noise	Riparian reserves are demarcated (buffers) to protect water ways. Bridges and culverts are installed to cross rivers/streams. Road width and canopy openings are minimized. Gravel is left in stream beds (for spawning). Reduced-impact logging (RIL) guidelines are strictly adhered to
Harvesting	Alteration of natural forest stand structure. Noise. Animals displaced from their natural habitat. Tree fall and shock. Loss of food supply. Habitat disturbance	Trees are marked. Directional felling is conducted. Trees that serve as seed sources, food sources and breeding niches for birds (trees > 120 cm dbh) are not felled or harvested. Roads are pre-aligned. Riparian reserves and buffer strips are maintained. Pockets of areas (>2 ha) above 250 within the compartment are mapped and excluded from harvesting. RIL guidelines are strictly adhered to
Silviculture	Elimination of woody vines that are a food source for some animals, especially birds, and also ladders for orangutans	Removal of immediate competitors only (non-commercial trees). Maintain structural diversity to encourage natural regeneration. Avoid use of chemical defoliators
Land clearing for agriculture using fire by villagers outside the reserve along common boundaries.	Forest fire. Complete annihilation of forest	Fire management plan. Fire crews. Fire fighting equipment. Fire preparedness plan. Fire prevention plan. Fire danger rating. Community services to build awareness
Hunting	Elimination of some endangered species. Forest fires	Installation of barrier/gate at main access road. Closing all known access leading into Deramakot Forest Reserve. Surveillance and patrols

et al. 2007). The former estimated the population of orangutans in Deramakot Forest Reserve as 792 and the density as 1.5 per km^2 in 1999 (Ancrenaz et al. 2005). The latter study demonstrated the importance of natural licks for the conservation of large mammals, where 80% of the medium to large mammals of Sabah were recorded (Table 4).

The previous study clearly pointed out the importance of the conservation of medium to large mammals in production forests of Deramakot. Routine activities are being conducted in Deramakot Forest Reserve to monitor the population dynamics of wildlife. One is the orangutan aerial nest count, which is conducted twice each year and the other is opportunistic sightings (on a daily basis). The aerial counts of orangutan nests help to monitor the "health" of orangutan populations (Mannan et al. 2003). If a sudden and significant drop in the number of nests occurs, deleterious actions could have affected orangutans and these deleterious factors need to be identified immediately. Daily opportunistic sightings also help to monitor the species distribution of the various animal species.

High conservation value forest

The Forest Stewardship Council's principles require the establishment of a certain area to be protected as an HCVF. We have set aside such an area in Deramakot Forest Reserve that satisfies all criteria of an HCVF and provides a key habitat for five globally threatened large mammals, namely the orangutan, Asian elephant, banteng, proboscis monkey, and clouded leopard. Large mammals need large areas to forage, and taking measures to conserve these areas would certainly help in protecting other smaller animals that occupy the same habitat. HCVFs, as defined in this plan, are the forest entities that possess one or more of the following attributes: (1) high biodiversity values (e.g., areas of high endemism, areas known to support endangered species, areas rich in wildlife), (2) rare, threatened or endangered ecosystems, (3) representative samples of natural populations in their undisturbed form (e.g., pristine forest), (4) provision of basic nature services in critical situations (e.g., watershed protection, erosion control), and (5) areas fundamental to meeting the basic needs of local communities (e.g., subsistence, proteins, medicines,

Table 4 Medium to large mammal fauna in the Deramakot Forest Reserve

Order	Family	Species (scientific name)
Insectivora	Erinaceidae	Moon rat (*Echinosorex gymnurus*)[a]
Primates	Lorisidae	Slow loris (*Nycticebus coucang*)
	Tarsiidae	Western tarsier (*Tarsius bancanus*)
	Cercopithecidae	Red leaf monkey (*Presbytis rubicunda*)
		Silvered langur (*Presbytis cristata*)[a]
		Proboscis monkey (*Nasalis larvatus*)
		Long-tailed macaque (*Macaca fascicularis*)
		Pig-tailed macaque (*Macaca nemestrina*)[a]
	Hylobatidae	Bornean gibbon (*Hylobates muelleri*)
	Pongidae	Orangutan (*Pongo pygmaeus*)[a]
Pholidota	Manidae	Pangolin (*Manis javanica*)[a]
Rodentia	Hystricidae	Long-tailed porcupine (*Trichys fasciculata*)[a]
		Common porcupine (*Hystrix brachyuran*)[a]
		Thick-spined porcupine (*Thecurus crassispinis*)[a]
Carnivora	Ursidae	Sun bear (*Helarctos malayanus*)[a]
	Mustelidae	Yellow-throated marten (*Martes flavigula*)
		Malay badger (*Mydaus javanensis*)[a]
		Oriental small-clawed otter (*Aonyx cinerea*)[a]
	Viverridae	Malay civet (*Viverra tangalunga*)[a]
		Otter-civet (*Cynogale bennettii*)[a]
		Binturong (*Arctictis binturong*)[a]
		Masked palm civet (*Paguma larvata*)[a]
		Common palm civet (*Paradoxurus hermaphroditus*)[a]
		Banded palm civet (*Hemigalus derbyanus*)[a]
		Short-tailed mongoose (*Herpestes brachyurus*)[a]
		Collared mongoose (*Herpestes semitorquatus*)[a]
	Felidae	Clouded leopard (*Neofelis nebulosa*)[a]
		Flat-headed cat (*Felis planiceps*)
		Leopard cat (*Felis bengalensis*)[a]
Proboscidea	Elephantidae	Asian elephant (*Elephas maximus*)[a]
Artiodactyla	Suidae	Bearded pig (*Sus barbatus*)[a]
	Tragulidae	Lesser mouse-deer (*Tragulus javanicus*)[a]
		Greater mouse-deer (*Tragulus napu*)[a]
	Cervidae	Bornean yellow muntjac (*Muntiacus atherodes*)[a]
		Red muntjac (*Muntiacus muntjak*)[a]
		Sambar deer (*Cervus unicolor*)[a]
	Bovidae	Banteng (*Bos javanicus*)[a]

[a] Species recorded at natural licks

building materials, and clean water) and/or are critical to local communities' cultural integrity (e.g., areas of cultural and ecological significance).

About 4,000 ha of forests (compartments) within Deramakot Forest Reserve that are steeply dissected (with slope gradient above 25°) have been permanently set aside for protection as HCVF. However, other areas for timber production may also contain HCVF where biological and ecological values are high. Therefore, Sabah Forestry Department will conduct a review of these protection areas. At present, some natural licks are under examination for inclusion into a HCVF because of the high dependence of endangered large mammals such as orangutan, Asian elephant, and banteng on the natural licks (Matsubayashi et al. 2007).

Silviculture and rehabilitation planting improve the forest ecosystem

Silviculture seeks to eliminate weed species (climbers, creepers, bamboo, etc.) that smother and suppress regrowth of desired species. The potential crop trees of the future, by and large, are the climax species that have evolved with the wildlife. Rehabilitation planting includes tree species with large fruits that became food sources for the primates found in Deramakot Forest Reserve (including the genera *Durio*, *Dracontomelon*, and *Mangifera*). Silvicultural activities are expected to maintain the food chain and ecosystem habitats to provide the food sources that wildlife depends on and to provide habitats for frugivorous insects.

Discussion

After some 15 years (1989–2005) of intensive management in Deramakot Forest Reserve, with 8 years under certification, what are the basic lessons that we have learned to make things better and to make things happen? Let us now consider the matters and issues taken from Deramakot Forest Reserve over more than one decade of trial and error.

Without political commitment from state leaders, the concept of Deramakot Forest Reserve could not have been expanded to other areas of Sabah and manifested in the long-term Sustainable Forest Management License Agreement policy launched in September 1997. Although the Sustainable Forest Management License Agreement arrangement is still in its infancy, it is a step in the right direction and far better than the previous ad hoc timber licensing system that previously prevailed, which could cause severe damage to the forest resources.

Forest certification was found to indirectly enhance log pricing. It serves as a catalyst for amendments to the timber marketing system by sorting species into user-oriented species groups. With a sense of perspective, we therefore consider the cost of certification as fair.

Conjecture about the benefits of certification to timber producers has centered on the "market premium" and "market access" debates. The market premium for logs is defined as the difference between the price of the certified log and the price of the same log prior to the adoption of certification (Varangis et al. 1995). The issue of whether or not certified logs fetch a market premium has been discussed controversially for years (Alstair 2002). Varangis et al. (1995) estimated that in view of the market share of certified tropical timber on the US and European markets, the incremental revenue from the markets assumed to be willing to pay more for certified timber would amount to 62 million USD. Our sales of logs by auction indicate that buyers do offer premium prices for certified logs by a margin of 51 USD per cubic meter which is equivalent to a price increase of 44% as compared to uncertified logs (Mannan et al. 2002; Kollert and Lagan 2005).

Reduced-impact logging and certification are also effective in wildlife management. It has been reported that the population of endangered large mammals, including orangutan, Asian elephant, and banteng, in Sabah has suffered from habitat loss, habitat fragmentation, and habitat degradation. Our own censuses and reports from other researchers demonstrate that Deramakot sustains denser populations of these animals. We conclude that good management of Deramakot can maintain the habitat contiguity of these animals. As these animals are known as umbrella species in the ecosystem, protecting them will add momentum to conserving entire ecosystem components.

Furthermore, studies on the floristics and the soil macrofauna showed that reduced-impact logging might have left lighter logging impacts on the forests than conventional methods did. For instance, tree species diversity was equally rich in the old-growth forest and in the forest harvested by reduced-impact logging, where climax and important commercial-timber species of Dipterocarpaceae dominated, but was much lower in the forest harvested by the conventional method, where pioneer species of the genus *Macaranga* (Euphorbiaceae) dominated (Seino et al. 2006). Moreover, the size structure of canopy-tree populations showed that dipterocarp trees regenerated well in the old-growth forest and the forest harvested by reduced-impact logging. By contrast, the pioneer species demonstrated rigorous regeneration in the forest harvested by the conventional method (Seino et al. 2006). The dominance of Dipterocarpaceae was related to the community structure of soil macrofauna, and this demonstrated that reduced-impact logging was also less destructive to soil macrofauna (Hasegawa et al. 2006).

Pristine forest ecosystems consist of plant–animal interactions in addition to physical environments and organisms. To achieve truly sustainable forest management on a longer time scale, it is necessary to create a management plan that incorporates such interactions. The strict application of reduced-impact logging and the establishment of an HCVF with an appropriate contiguity and area within a commercial forest reserve are definitely two effective measures to protect the physical environment, flora, fauna and interactions while producing timber in the reserve.

We hope that the sustainable management system of Deramakot Forest Reserve will be adopted as a model in other commercial forest reserves.

References

Alstair S (2002) Editorial. Tropical forest update. ITTO 12(2):1–2

Ancrenaz M, Gimenez O, Ambu L, Ancrenaz K, Andau P, Goossens B, Payne J, Sawang A, Tuuga A, Lackman-Ancrenaz I (2005) Aerial surveys give new estimates for orangutans in Sabah, Malaysia. PLOS Biol 3:1–8

Forestry Department of Sabah (1995) Forest management plan, Deramakot Forest Reserve (1.1.1995–31.12.2004). Forestry Department of Sabah, Sandakan, Malaysia

Forestry Department of Sabah (2005) 2nd Forest Management Plan, Deramakot Forest Reserve (1.1.2005–31.12.2014). Forestry Department of Sabah, Sandakan, Malaysia

Hasegawa M, Ito MT, Kitayama K, Seino T, Chung AYC (2006) Logging effects on soil macrofauna in the rain forests of Deramakot Forest Reserve, Sabah, Malaysia. In: Lee YF et al. (eds) Synergy between carbon management and biodiversity conservation in tropical rain forests. Proceedings of the 2nd workshop, Sandakan, Malaysia, 30 November–1 December 2005. DIWPA, Shiga, Japan

Huth A, Ditzer T (2004) Long-term impacts of logging in a tropical rain forest—a simulation study. For Ecol Manage 142:33–51

Kleine M, Heuveldop J (1993) A management planning concept for sustained yield of tropical forests in Sabah, Malaysia. For Ecol Manage 61:277–297

Kollert W, Lagan P (2005) Do certified tropical logs fetch a market premium? A comparative price analysis from Sabah, Malaysia. XXII IUFRO World Congress 2005, Brisbane, Australia

Lee YF, Chung AYC, Kitayama K (2006) Synergy between carbon management and biodiversity conservation in tropical rain forests. Proceedings of the 2nd workshop, Sandakan, Malaysia, 30 November–1 December 2005. DIWPA, Shiga, Japan

Mannan S, Awang Y, Radin A, Abi A, Lagan P (2002) The Sabah Forestry Department experience from Deramakot Forest Reserve: 5 years of experience in certified sustainable forest management. UMS Seminar, August 2002

Mannan S, Awang Y, Radin A, Abi A, Suparlan SH, Lagan P (2003) Conservation of the orangutan and the forest management units: the Deramakot model. Workshop on orangutan conservation in Sabah, August 2003

Matsubayashi H, Lagan P, Majalap N, Tangah J, Sukor JRA, Kitayama K (2007) Importance of natural licks for the mammals in Bornean inland tropical rain forests. Ecol Res (in press)

Ong RC, Kleine M (1995) DIPSIM: a dipterocarp forest growth simulation model for Sabah. Forest Research Centre, Sabah Forestry Department, Sandakan, Malaysia

Seino T, Takyu M, Aiba S, Kitayama K, Ong RC (2006) Floristic composition, stand structure, and above-ground biomass of the tropical rain forests of Deramakot and Tangkulap Forest Reserve in Malaysia under different forest managements. In: Lee YF et al. (eds) Synergy between carbon management and biodiversity conservation in tropical rain forests. Proceedings of the 2nd workshop, Sandakan, Malaysia, 30 November–1 December 2005. DIWPA, Shiga, Japan

Varangis P, Crossley R, Braga C (1995) Is there a commercial case for tropical timber certification? World Bank policy research working paper 1479. World Bank, International Economics Department, Commodity Policy and Analysis Unit, Washington DC

Ecol Res (2007) 22: 422–430
DOI 10.1007/s11284-007-0363-2

SPECIAL FEATURE

Sustainability and biodiversity of forest ecosystems:
an interdisciplinary approach

Ken-Ichi Akao · Y. Hossein Farzin

When is it optimal to exhaust a resource in a finite time?

Received: 27 January 2006 / Accepted: 27 November 2006 / Published online: 3 April 2007
© The Ecological Society of Japan 2007

Abstract Exhaustion of a natural resource stock may be
a rational choice for an individual and/or a community,
even if a sustainable use for the resource is feasible and
the resource users are farsighted and well informed on
the ecosystem. We identify conditions under which it is
optimal not to sustain resource use. These conditions
concern the discounting of future benefits, instability of
a social system or ecosystem, nonconvexity of natural
growth function, socio-psychological value of employment, and strategic interaction among resource users.
The identification of these conditions can help design
policies to prevent unsustainable patterns of resource
use.

Keywords Renewable resource management ·
Sustainability · Finite-time exhaustion ·
Optimal path · Policy implications

JEL Classification Code Q01 · Q20

Introduction

Sustainability has long been a primary objective of
renewable resource management. The notion of sustained yield goes back to at least eighteenth century
European forestry (Carlowitz 1713. Also see Vanclay
1996). After the publication of "Our Common Future"
(World Commission on Environment and Development
1987) and the United Nation Conference on Environ-

K.-I. Akao (✉)
School of Social Sciences, Waseda University,
1-6-1 Nishi-waseda, Shinjuku, Tokyo 169-8050, Japan
E-mail: akao@waseda.jp
Tel.: +81-3-52861908
Fax: +81-3-32048962

Y. H. Farzin
Department of Agricultural and Resource Economics,
University of California, Davis, One Shields Avenue,
Davis CA95616, USA
E-mail: farzin@primal.ucdavis.edu

ment and Development in Rio de Janeiro, 1992, the
concept of sustainability has been popularized and regarded as one of the basic social goals. At the same time,
most scholars have recognized the vagueness of the
concept, raising questions such as: is it to keep physically
intact a natural resource or environmental asset? If so,
how should one think about the sustainability of a
nonrenewable resource? How does it relate to intergenerational equity and intertemporal efficiency? Not sur-
prisingly, economists have come to various definitions of
sustainability and differing views about their merits (see,
for example, Pearce et al. 1990; Turner et al. 1994;
Nordhaus 1994; Solow 1998; Heal 1998; Farzin 2004).

Whatever the definition of sustainability, it is obvious
that finite-time resource extinction defies sustainability.
In this paper, we show how a rational agent willingly
exhausts a resource in a finite time, even though a sustainable resource use is feasible, or, at least, the resource
could be used up over an infinitely long period. The
assumption of rationality is important: it enables us to
avoid an unsustainable path of resource use by removing
the very conditions that render finite-time extinction
rational. Therefore, the aim of the paper is to identify
the conditions under which finite-time exhaustion of a
renewable resource is optimal. These conditions concern
(1) the discounting of future benefits, (2) uncertainty
about the future of the resource stock, (3) nonconvexity
of natural growth function, (4) socio-psychological aspect of work incentives, and (5) strategic interaction
among resource users.

The paper is organized as follows: the next section
introduces a simple model for resource management to
show that heavy discounting makes finite-time extinction
optimal. We show that a source of a high discount rate is
the uncertainty about the future ecological state of the
resource stock or about its future ownership and management. In Sect. 3, we modify the model by allowing
nonconvexity in the resource's natural growth function. If
an inbreeding depression or an Alee effect exists, the
growth function takes a shape that it is convex when the
population size (resource stock) is small and concave

423

when it is large. We will see that even with a low discount rate, if the initial stock of the resource is small, the optimal resource policy is finite-time exhaustion. In Sect. 4, the model is extended to incorporate the socio-psychological value of employment. We show that even with a low discount rate and an abundant resource stock, finite-time exhaustion becomes optimal. This is because in this case it is optimal for the resource user to harvest the resource at the maximum harvesting ability. This is an extreme case of extinction: the most rapid extinction. In Sect. 5, we consider a common property resource problem, assuming that multiple agents use the resource. Again, we show that the most rapid extinction is optimal for each individual resource user. At the same time, we show that, under the same condition, sustainable resource use is an optimal policy, too. However, one cannot be sure which optimal policy is adopted. Section 6 concludes with some policy implications of these findings.

Discounting and uncertainty

Let us start with a rudimentary model in resource economics, characterized by the following problem:

$$\max_{c(t) \geqslant 0} \int_0^\infty u(c(t))e^{-\rho t}dt \tag{1a}$$

subject to $\dot{x}(t) = f(x(t)) - c(t),$ (1b)

$x(t), c(t) \geqslant 0, x(0) = x_0$ given. (1c)

Here x denotes the stock of a renewable resource. The natural growth of the resource is described by function $f(x)$. Variable $c(t)$ denotes the amount of harvest at time t. Therefore, the evolution of the resource stock is described by Eq. (1b), where $\dot{x}(t)$ denotes the time derivative of $x(t)$. The consumption of harvest yields utility to the resource user according to the utility function $u(c)$. We assume that the natural growth function f is hump-shaped and strictly concave, and the utility function u is bounded from below, strictly increasing and strictly concave. See Fig. 1. Formally, we make the following assumptions:

Assumption 1 $f(x)$ satisfies $f(0) = f(K) = 0, K > 0, f'' < 0.$

Assumption 2 $u(c)$ satisfies $u(0) = 0, u' > 0$ and $u'' < 0.$

The objective of the resource user is to maximize the sum of his discounted utilities from the present time to infinite future, as seen in the objective functional (1a). The discount rate ρ is the user's time preference parameter. If ρ is zero, the user values the utilities equally between now and any time in the future, whereas if it is positive, the utilities in the future are valued less than the present utility. In particular, by the power of discounting, the present value of the well-being of a generation living in a far distant future is almost negligible. This implies that the choice of a discount rate raises an ethical problem for intergenerational equity. In fact, in the seminal paper that initiated dynamic analysis in economics, Frank Ramsey, who was a philosopher and mathematician as well as economist, wrote that (discounting) is ethically indefensible and arises merely from the weakness of the imagination (Ramsey 1928). It should be noted, however, that discounting could be rationalized from a non-ethical standpoint. For now, let us assume a positive discount rate $\rho > 0$. Later, we will justify it for a non-ethical reason: uncertainty.

In order to obtain finite-time extinction as an optimal path, we make an additional assumption:

Assumption 3 $f'(0) = r < \infty$ and $u'(0) < \infty.$

The first inequality implies that the intrinsic (biological/natural) growth is finite no matter how small the resource stock may be, which is a quite plausible assumption for any renewable resource. The second inequality implies that the marginal value of the resource is finite no matter how scarce it may become, which is also a plausible assumption except in extreme cases where the resource is absolutely vital to life, as would be the case, for example, with water or oxygen. In these cases, if your consumption of water or oxygen shrinks to zero, the value of a unit of them for you, $u'(0)$, will rise up to infinity. Therefore, you will never exhaust the

Fig. 1 Natural growth function and utility function

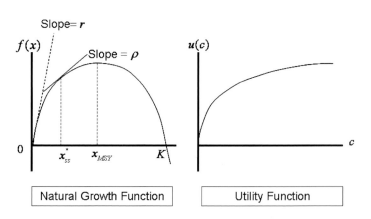

Natural Growth Function

Utility Function

source of water or oxygen. This explains the necessity of this assumption for optimal finite-time resource exhaustion. However, in a setup with multiple resource users, as in Sect. 5, the finite-time resource extinction could be optimal even if the resource is essential.

To solve the rudimentary problem, it is routine to define the Hamiltonian as

$$H(c, x, \lambda) = u(c) + \lambda(f(x) - c), \quad (2)$$

where λ is called the costate variable or the shadow price of the resource stock. By Pontryagin's maximum principle, there is a function of time $\lambda(t)$, for which an interior optimal control $c^*(t)$ and the optimal state $x^*(t)$ should satisfy

$$\frac{\partial H(c^*, x^*, \lambda)}{\partial c} = u'(c^*) - \lambda = 0, \quad (3a)$$

$$\dot{\lambda} = \rho\lambda - \frac{\partial H(c^*, x^*, \lambda)}{\partial x} = \lambda[\rho - f'(x^*)], \quad (3b)$$

for each $t \geq 0$.

From (1b) and (3b), an interior optimal steady state (denoting by subscript "ss") satisfies

$$f(x_{ss}^*) - c_{ss}^* = 0 \quad \text{and} \quad f'(x_{ss}^*) = \rho. \quad (4)$$

An optimal steady state (shortly, OSS) is the level of resource stock that once attained, it is optimal to sustain the level. For our model, it can be shown that if an interior OSS $x_{ss}^* > 0$ exists, it is unique and every optimal path starting from any initial stock level monotonically converges to it (see Fig. 2a). The uniqueness follows from the strict concavity of f (for the monotonicity and stability property, see Hartl 1987; Levhari and Liviatan 1972, respectively). A long-run target for resource policy is, thus, to tend the interior OSS, and the resource policy is sustainable. However, for an interior OSS to exist, the following necessary condition has to be satisfied: the intrinsic growth rate exceeds the discount rate:

$$\rho < r. \quad (5)$$

Roughly speaking, this condition requires that the harvesters should not be 'too' impatient to make it worthwhile to extract the resource at a rate faster than the rate at which it is capable to regenerate itself. Otherwise, it

would lead to the resource's eventual extinction. In other words, if the discount rate is too high to satisfy inequality (5), the interior OSS would not exist. In this case, the zero stock level ($x = 0$) is the only OSS and an optimal path starting from any initial resource stock, *no matter how large*, converges to this level. See Fig. 2b for the typical optimal paths. Furthermore, as proved in the Appendix of this paper, under Assumption 3, it takes only a finite time for the optimal policy to reach this level. The resource policy is unsustainable:

Proposition 1 *For the rudimentary problem* (1) *with Assumptions* 1–3, *if the discount rate exceeds the intrinsic growth rate* ($\rho > r$), *finite−time extinction is optimal for the resource user no matter how large the initial resource stock size is.*

The economic intuition for this proposition is simple: because of the assumption of the concavity of the natural growth function, $f''(x) < 0$ and the definition of $r = f'(0)$, the condition that the discount rate ρ exceeds r implies that *at all levels* of resource stock, investing in the resource stock (resource conservation) yields a rate of return $f'(x)$ less than the opportunity cost of that investment ρ, thus giving the users the incentive not to invest in the resource stock, which means resource depletion to extinction.

As mentioned before, the choice of discount rate raises an ethical problem: that is, how should we value the well-being of future generations? Another problem is that if social preferences are inherent such that they imply too high a social discount rate, thus causing unsustainable resource use, then there is little that one can do to prevent extinction without resorting to regulation of resource harvesting in some fashion. However, independently of any ethical argument, there is another reason for discounting: that is, uncertainty about the future of the resource stock or its ownership. Imagine that a sudden disaster completely destroys the resource, or the resource owner is suddenly deprived of his ownership by, say, confiscation of the resource stock by a corrupt or politically radical government. Note that for environmental conservation or other reasons, a politically radical government may suddenly impose a resource tax whose rate is sufficiently high and/or increases

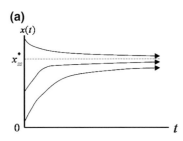
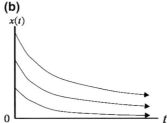
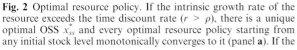

Fig. 2 Optimal resource policy. If the intrinsic growth rate of the resource exceeds the time discount rate ($r > \rho$), there is a unique optimal OSS x_{ss}^* and every optimal resource policy starting from any initial stock level monotonically converges to it (panel **a**). If the

growth rates is less than or equal to the time discount rate ($r \leq \rho$), every optimal resource policy is resource exhaustion. In particular, if $r < \rho$ and Assumption 3 are satisfied, the optimal extinction occurs in a finite time steadily at a constant proportional rate. It will have the equivalent effect as a stochastic resource confiscation/catastrophe, and as shown here, it will cause finite-time exhaustion to be optimal.

Suppose that this sort of fatal event occurs with a positive probability. Formally, we assume that the agent does not discount future utilities at all. Instead, the parameter ρ expresses the hazard rate of the Poisson process for the occurrences of the fatal event. Once the fatal event occurs, the utility levels of the resource user thereafter are zero forever. Note that the probability with which the event occurs within the interval of time $[t, t + \mathrm{d}t)$ is $\rho e^{-\rho t}\mathrm{d}t$. Then, objective functional (1a) is modified as

$$
E\left[\int_0^T u(c(t))\mathrm{d}t\right] = \int_0^\infty \left[\int_0^T u(c(t))\mathrm{d}t\right]\rho e^{-\rho T}\mathrm{d}T
$$
$$
= \int_0^\infty u(c(t))e^{-\rho t}\mathrm{d}t. \tag{6}
$$

So, we are back to the rudimentary problem (1a, 1b, 1c), although now ρ expresses the magnitude of the probability of the fatal event, and not the discount rate. We interpret this result as the following corollary:

Corollary of Proposition 1 *Finite−time extinction may be optimal if the ecosystem and/or the socio−political system is so unstable that the probability of the arrival of the ecological catastrophe or socio−political upheaval is so high as to exceed the intrinsic growth rate, $\rho > r$.*

Remark 1 For the derivation of (6), see, for example, Dasgupta and Heal (1979). Recent models of optimization under uncertainty show that the hazard rate can be viewed as an effective discount rate even when it is not a constant parameter and when the event is not fully fatal (i.e., the post-occurrence welfare does not vanish). See Tsur and Zemel (2006) and the literature therein cited.

Nonconvexity of a natural growth function

In this section, we focus on the natural growth function. A concave growth function implies that the natural growth rates increase as the size of the resource stock decreases ($f''(x) < 0$). However, as is well known in population biology literature, if the stock size is very small, the growth rate may be small, for example, due to an Alee effect or an inbreeding depression. Then, we may have a convex–concave shape of the growth function, as in Fig. 3.

Now, consider the rudimentary problem (1a) with modification of Assumption 1 as follows:

Assumption 4 $f(0) = 0, f'(0) < \rho, \exists\ x_I > 0: f'(x_I) > \rho,$ $f''(x) > (<)0$ if $x < (>)x_I.$

Notice that because of the convexity of $f(x)$ on $[0, x_I]$, the condition $f'(0) < \rho$ is a *milder* assumption than the corresponding one in the previous case of concave growth function in that discount rate ρ need no longer be greater than the growth rate $f'(x)$ at all stock levels. In fact, it may be *less than* that for some stock levels. There are two stock levels at which the slope of the growth function equals to the discount rate, shown in Fig. 3 at the tangency points. The larger one x_{ss}^* corresponds to the OSS for the original rudimentary model (1a). The smaller one, x_ρ, is new. However, it can be shown that x_ρ is not an OSS (see Akao and Farzin 2006). Therefore, we have potentially two optimal paths: one converges to the interior optimal sustainable resource stock level and the other converges to zero stock level, i.e., extinction in a finite time. Then, we have the following proposition:

Proposition 2 *Suppose the existence of optimal paths. Under Assumption 4, there exists a threshold $x_c \in (0,\infty]$ such that if $x_0 < (>)x_c$, the optimal path monotonically converges to $0(x_{ss}^*)$. Furthermore, the threshold satisfies $x_c < x_{ss}^*$, if the following mild discounting condition holds:*

$$\rho < \max[f(x)/x | x \geqslant 0].$$

Proof See Akao and Farzin (2006).

Notice that compared to Proposition 1, while the convex–concavity of the growth function modifies the requirement that the discount rate should exceed the growth rate at all stock levels in order for the finite time extinction to be optimal, it also weakens Proposition 1 in that a finite time extinction is optimal only if the initial resource stock is less than a critical level.

Figure 4 illustrates the optimal paths. Even with a very low discount rate, in the presence of non-concavity of natural growth function, finite-time extinction may be optimal if the resource stock has been already degraded (by, for example, overexploitation thus far) below the critical threshold. This threshold x_c is called the Skiba point or DNS point.

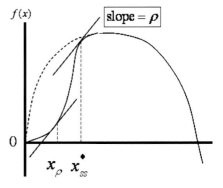

Fig. 3 Convex–concave natural growth function

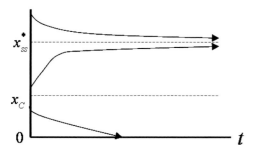

Fig. 4 Optimal paths when the natural growth function is convex–concave. In the case with a convex–concave natural growth function, there is a critical stock level of the resource x_C. If the initial stock is higher than the critical level, the optimal path converges to the interior steady state, so that the optimal resource policy is sustainable. If the initial stock is lower than the critical level, the optimal path goes to zero stock level in a finite time, so that the optimal resource policy is unsustainable

Remark 2 If the reverse inequality holds in Proposition 2, then the interior OSS disappears, so that despite the convex–concave condition of the growth function the only optimal policy is extinction in a finite time starting from any initial stock level.

Remark 3 "DNS" are the initials of three authors, Dechert, Nishimura and Skiba. Skiba (1978) first introduced the convex-concave production function in the theory of optimal growth in economics. A rigorous analysis for a discrete time model is given by Dechert and Nishimura (1983). Gustav Feichtinger and his collaborators have recently studied the continuous time models and their applications. See, for example, Deissenberg et al. (2001) and Hartl et al. (2004), which contain the literature review in economic dynamics with nonconvexity, including environmental economics.

Non-pecuniary value of employment: a socio-psychological aspect

It is natural to think that working is not only a means of earning income, but also a form of social involvement. Because of this, unemployment usually brings a person unhappiness more than loss of income does, which may include, for example, losses of dignity, confidence and identity. In other words, there is a non-pecuniary value of employment. Curiously enough, this fact has been ignored in traditional economics until recently. Farzin and Akao (2006) incorporate this socio-psychological aspect explicitly into a bio-economic model and find that the optimal resource use may be finite-time extinction. In this section, we introduce their results with a simpler model than theirs.

We modify the utility part of the rudimentary problem (1a) as follows:

Assumption 5 $u(c, E), E \in [0, \bar{E}], E = $ working time. $\partial u / \partial E > 0, \partial u^2 / \partial E^2 < 0$.

The utility stems from two sources. One is consumption of harvests as before. The other source is working. Different from a standard economic model, here working is not a source of disutility, but a source of utility. Although this assumption may seem curious, Farzin and Akao (2004) show that under standard assumptions in economics, non-pecuniary value of work exceeds the value of leisure at very low income levels.

Assume that all the labor is used to extract the resource, which is the case where there is no alternative employment other than resource extraction. The relationship between labor input E and resource output c is described with the cost function $E(c)$. There is an upper bound for working time $\bar{E}$, which limits the maximum harvest level. Denote the maximum harvest level by $\bar{H}$, which satisfies $\bar{E} = E(\bar{H})$. We assume that with the maximum effort $\bar{E}$, the resource is certainly exhausted in a finite time:

Assumption 6 The maximum harvest level with full employment exceeds the maximum sustained yield (MSY): $\bar{H} > f(x_{MSY})$, where x_{MSY} is the stock level yielding of the maximum sustainable harvest.

Remark 4 See Fig. 1 for the geography of the stock of maximum sustained yield, x_{MSY}, which in general does not coincide to the interior OSS, x_{ss}^*, although MSY has been a long-run target of natural resource management. Clark (1976) discusses these two concepts at some length.

It is important to notice that even though working is a source of utility, the full employment $\bar{E}$ is *not necessarily* an optimal choice. This is because the full employment may degrade the resource too much to allow sustaining future consumptions. Recall that we have supposed that the resource user is rational enough and in particular farsighted.

Although it is mathematically invariant, let us add a flavor of macro economics to the rudimentary model (1). Consider a community, in which the local people are governed by a benevolent government. Everyone has identical preferences and the same harvesting technology, as described above. Let n be the population size. The problem of the benevolent government is:

$$\max_{c(t) \geqslant 0} \int_0^\infty u[c(t), E(c(t))] \mathrm{e}^{-\rho t} \mathrm{d}t \tag{7a}$$

subject to $\dot{x}(t) = f(x(t)) - nc(t),$ (7b)

$$0 \leqslant E(c) \leqslant \bar{H}, x(0) = x_0 \text{ given.} \tag{7c}$$

Pontryagin's maximum principle suggests that an optimal control c^* maximizes the Hamiltonian:

$$H(c, x, \lambda) = U(c) + \lambda[f(x) - nc],$$

where $U(c) = u[c, E(c)]$.

Assume that the reduced form utility function U is strictly *convex* in c. The following example shows that such a convex utility function is obtained with standard assumptions in economics, if we allow working to be a source of utility.

Example Let utility function have a form of $u(c,E) = c^\alpha E^\eta$, with $0 < \alpha < 1$ and $0 < \eta < 1 - \alpha$ (so that u is increasing and concave jointly in c and E, a standard assumption of economics). The harvesting technology is expressed by $E(c) = c^\beta$ with $\beta > 1$, which is also a standard assumption of economics: a cost function is convex and increasing. If the elasticity of marginal utility of employment is sufficiently high to satisfy $\beta > (1 - \alpha)/\eta$, then $U(c) = u[c, E(c)] = c^{\alpha + \beta\eta}$ is strictly convex ($\mathrm{d}^2 U/\mathrm{d}c^2 > 0$).

If $\mathrm{d}^2 U/\mathrm{d}c^2 > 0$, the maximum of the Hamiltonian is attained at a corner of c. That is, the optimal control c^* is either of the full harvesting $c* = \bar{H}$ or no harvesting $c* = 0$. Also, notice that there is no interior optimal control and thus no interior OSS. Therefore, the optimal path of the resource stock converges either to the carrying capacity or to zero. It is, however, obvious that the path going to the carrying capacity is suboptimal, because there is no chance to harvest at all. Therefore, we have:

Proposition 3 *Under Assumptions 5 and 6, if* $\mathrm{d}^2 u[c,E(c)]/\mathrm{d}c^2 > 0$, *full employment is always optimal. On an optimal path the resource stock decreases most rapidly and becomes extinct in a finite time.*

(The formal proof, including the existence of an optimal path, is found in Farzin and Akao 2006.)

If the harvest level with full employment exceeds the MSY, full employment and sustainable resource management are incompatible objectives, and in Farzin and Akao's framework, the former is chosen over the latter as the optimal policy. Population growth and technological progress in resource extraction may bring about such a situation. The optimal path has two novelties. First, the optimal resource extinction is an extreme one, the most rapid extinction. Second, resource extinction is optimal irrespective of the state of the resource stock and the magnitude of the discount rate. Notice that in this section, we have referred neither to the discount rate nor to the initial level of the resource stock, which were crucial factors for finite-time extinction to be optimal in the previous sections.

Remark 5 Without invoking non-pecuniary value of employment, we could obtain the most rapid extinction as an optimal path. It is necessary, however, to specify the utility and natural growth functions that satisfy the restrictive conditions derived by Spence and Starrett (1975). Heavy discounting is also needed.

Strategic interaction

In this section, we consider a natural resource used by multiple users. Such a resource may be categorized by its physical property into two types. The first type is a resource for which it is difficult to establish and force a definite property right. The global atmosphere, underground aquifers, and highly migratory fish stocks are a few examples. The second type is a resource that, although its private or governmental holding is physically possible, is owned communally for institutional or historical reasons. An example is the high seas defined in the United Nations Convention on the Law of the Sea. Another example is a local communal forest in Japan, which is a relic of the Edo era (1603–1868), at which time private ownership of a forest was prohibited.

We will show that for those resources, finite-time extinction may be optimal from the viewpoint of each rational user, despite the fact that it is by no means socially or cooperatively optimal. In other words, we will see the individual optimality of the so-called "tragedy of the commons." However, it turns out that finite-time extinction is not the only non-cooperative optimal path and under certain conditions a sustainable resource use is optimal, too. Therefore, the tragedy of the commons is not an inevitable destiny. This could explain the fact that some communal resources have been managed in a sustainable way, at least apparently and so far.

A fundamental change from the previous models is that not a single agent, but many agents use the resource. We assume that the number of resource users $n \geq 2$ is fixed. In the terminology of economics, this sort of resource is called a common property resource or a common pool resource. The resource users are identical in their preferences and harvesting technology. As in the previous section, there exists the upper bound of the harvest ability $\bar{h}(= \bar{H}/n) > 0$. Modifying the rudimentary model (1), we study the following differential game model:

$$\max_{c(t) \geqslant 0} \int_0^\infty u(c(t))\mathrm{e}^{-\rho t}\mathrm{d}t \tag{8a}$$

subject to $\dot{x}(t) = f(x(t)) - (n-1)\sigma(x) - c(t),$ (8b)

$c(t) \in [0, \bar{h}], x(0) = x_0$ given. (8c)

where $\sigma(x)$ is the common harvesting strategy of all other users, which is assumed to depend only on the common pool resource stock, and in particular, to be time independent. An example of such a strategy is the most rapid extinction strategy defined as below:

$$\sigma(x) = \begin{cases} \bar{h} \\ 0 \end{cases} \text{ if } \begin{matrix} x > 0 \\ x = 0 \end{matrix}, \tag{9}$$

which harvests with the maximum effort as far as the resource exists. When the other players use the same strategy $\sigma(x)$, the problem for each player is written as in (8a, 8b, 8c) above. If the optimal solution is described by the same strategy $\sigma(x)$, $\sigma(x)$ constitutes a symmetric Nash equilibrium. It is a Nash equilibrium because once all players choose their strategies, then no one wants to deviate from it. Such equilibrium also has the property that even if some players deviate from the equilibrium strategy at some point in time, the equilibrium strategy will be still optimal if at later time the players return to that equilibrium strategy. This property in economics is referred to as strongly time consistent or as "subgame perfect." Finally, it is symmetric since all players adopt the same strategy. For analytical simplicity, we will restrict our concern to this type of equilibrium, i.e., symmetric Markov perfect Nash equilibrium.

Notice that the problem of each resource user now becomes more complicated than the ones in the previous sections, because other users also harvest the resource and their harvest rates affect the user's action, which affects, in turn, other users' actions. This is the strategic interaction.

The cooperative or social optimization problem, compared with non-cooperative problem (8a, 8b, 8c), is formulated as follows:

$$\max_{c(t) \geqslant 0} \int_0^\infty u(c(t)) e^{-\rho t} dt$$

subject to $\dot{x}(t) = f(x(t)) - nc(t)$,

$$c(t) \in [0, \bar{h}], x(0) = x_0 \quad \text{given}.$$

Maintain Assumptions 1, 2, and

Assumption 7 $f'(0) > \rho$.

We deliberately make this assumption to sharply distinguish the roll of strategic interaction in generating the optimality of finite time extinction from that of the high discount rate studied in Sect. 2. The cooperative problem above has the same property as the rudimentary problem (1) in mathematical terms. To see this, define the aggregate harvests by $C(t) = nc(t)$ and rewrite the instantaneous utility as a function of $C(t)$: $U(C(t)) = u(C(t)/n)$. Finally, verify that $U(C)$ satisfies Assumption 2. Therefore, every optimal cooperative path of the resource stock monotonically converges to a unique social OSS, $x_{ss}^* > 0$, such that $f'(x_{ss}^*) = \rho$. The cooperative resource policy is sustainable.

We want to show that finite-time exhaustion is a Nash equilibrium, i.e., individuals' rational choice. To do so, unlike the previous section, we do not need Assumption 3 (the finiteness of the marginal utility $u'(0) < \infty$ and the marginal productivity $f'(0) < \infty$ at the origin). Instead, we assume

Assumption 8 $n\bar{h} > f(x_{MSY})$, $(n-1)\bar{h} > f(x_{ss}^*)$, $\beta(c) = \frac{-cu''(c)}{u'(c)}$ $\leqslant \frac{n-1}{n}$, $\beta(0) > 0$.

The first inequality is the same assumption as Assumption 6 in the previous section. That is, with the maximum harvesting effort, the resource is exhausted in a finite time. The second inequality ensures that if other users harvest the resource with the maximum effort, there is no way for an individual user to sustain the social OSS, x_{ss}^*, because simply it is infeasible. The third

and forth inequalities restrict the curvature of utility function. These assumptions are technical, but standard in economics.

The following proposition on equilibrium resource use is given by Sorger (1998).

Proposition 4 (Sorger1998)

(a) *The most rapid extinction strategy* (9) *constitutes a Nash equilibrium if and only if the following inequality holds*:

$$u'(\bar{h}) \geqslant \frac{u(\bar{h})}{n\bar{h} - f(x_{ss}^*)} \exp\left[-\rho \int_0^{x_{ss}^*} \frac{dy}{n\bar{h} - f(y)}\right]. \quad (10)$$

(b) *There is a continuum of the other Nash strategies if* $f(x_{ss}^*) > nf'(x_{ss}^*)x_{ss}^*$. *Each strategy is sustainable in the sense that the associated equilibrium path of the resource stock converges to a positive stock level in* $(0, x_{ss}^*)$.

We refer to the strategy in Proposition 4(b) as Sorger's strategy, which is sustainable and under certain conditions can almost replicate the social optimum, although it always yields a payoff to each individual user which is less than what they would gain under the social optimum. In contrast, the most rapid extinction is the worst strategy, from the viewpoint of sustainability. The inequality in Proposition 4(a) holds when the number of users n is large, the maximum harvest rate $\bar{h}$ is high, or the discount rate of each user ρ is high. In any of these cases, finite-time extinction is optimal from an individual's viewpoint.

A troublesome, but interesting, point is that the most rapid extinction may coexist with Sorger's strategies as an equilibrium strategy and we cannot predict which equilibrium arises. This is illustrated in Fig. 5 with the following specification: the utility function is isoelastic, the natural growth function is described by a logistic equation, and ρ is fixed at a rather low level, since we have already seen how heavy discounting brings finite-time extinction. In Fig. 5, there are two areas: one is the area on which Sorger's strategies constitute equilibria and the other is the area on which the most rapid extinction strategy becomes an equilibrium strategy. Observe that these two areas overlap.

The coexistence of two types of equilibria means that the individually optimal resource policy for a common property resource is ambiguous and unstable. It is possible on a theoretical ground, that for two communities with identical resource stock and individual preferences, one uses its natural resource in a sustainable way, whereas the other exhausts its resource at the most rapid speed to leading to finite time extinction. The interest in this result is its potential to explain the observation that the resource based communities with similar economic

429

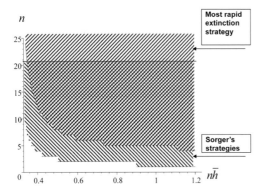

Fig. 5 Coexistence of sustainable and unsustainable equilibria. Depending on the maximum aggregate ability of harvesting $n\bar{h}$ and the number of users n, both two strategies may constitute their own Nash equilibria. Coexistence of two equilibria shown in the overlapped area implies the ambiguity and instability of common property resource management. This numerical example uses $u(c) = c^{0.3}$, $\rho = 0.03$, and $f(x) = 1.2x(1-x)$

and resource conditions have experienced the different resource management outcome, some are successful in sustaining their resources, while the others have depleted them to extinction. Another interesting implication of this theoretical result is that it can explain the possibility that a community which has been using its resource in a sustainable way for a period of time may suddenly switch to a ruinous resource use path without any evident trigger.

Remark 6 Another resource modeling with multiple users is that of an *open access resource problem*, for which the number of the users varies. Anonymous agents freely enter to extract the resource, until their temporal profits equilibrate to zero. For this setup, the mechanism causing finite-time extinction is quite simple. That is, it occurs if the aggregate harvest levels corresponding to zero profit always exceed the natural growth rates of the resource (Berck 1979). Also, notice that an open access resource problem can be regarded as the limit case of the problem (1) with $\rho \to \infty$ (Beddington et al. 1975). Therefore, the results in the second section are applied.

Concluding remarks

We conclude these observations with their policy implications.

First, we have seen how uncertainty raises discount rate and how a high discount rate brings finite-time resource extinction. To prevent such a situation, we need to mitigate the risk of the fatal events. For example, political stability matters. Furthermore, such a policy should be implemented early on, if the growth function of the resource exhibits nonconvexity and the resource is being degraded. This is because when the resource has

been already degraded, finite-time extinction is more prone to be an optimal resource use policy even with a low discount rate.

Second, we have seen that non-pecuniary value of employment makes people give priority to full employment over sustainable resource use. Farzin and Akao (2006) show that the remedy is none but to create alternative employment sources to absorb labor force which is excessive from the viewpoint of sustainable resource use. They also suggest that earlier policy implementation is more prudent, since when the resource is more degraded, higher wage rates may be necessary to prevent resource exhaustion.

Third, we have seen a strategic interaction cause the most rapid extinction, which is an extreme case of finite-time resource extinction. Although all common property resources do not have such a fate as predicted by Hardin (1968), all of them share the possibility. Breaking such an interaction is the primal policy to prevent the most rapid extinction. Akao (2001, 2004) shows that among standard economic policy measures, a tax on harvest does not work well, whereas tradable permits or quota system works to realize sustainable and efficient resource usage. Another prospective prescription is privatization. However, a caution is given by Dasgupta and Mäler (1997). They have pointed out that, in the real world, the consequence of privatization of a common property resource may be further resource degradation. This is due to the existing inequality in a rural community. If the resource is not favorably distributed to the poor, they cannot help but to encroach on the resource.

Finally, resource-sector technological assistance and income assistance may not help to prevent finite-time extinction. In particular, if a technological assistance improves the harvesting efficiency, and hence the maximum harvesting ability, it may even accelerate resource extinction.

Acknowledgments This paper was previously presented under the title of "Emergence of finite time resource extinction in resource economics" at the RIHN Pre-Symposium "Sustainability and Biodiversity of Forest Ecosystems—Drivers, mechanisms, and effects of forest change," held at Kyoto on October 18, 20onuma,05. The authors thank Amos Zemel, Ayumi Onuma, the late Kuniyasu Momose, Masahiro Ichikawa, Toru Nakashizuka, and other participants for useful comments.

Appendix: Proof of the optimal finite time extinction under Assumption 3

Assumptions 1 and 2 for the utility function and the natural growth function ensure the existence and uniqueness of optimal path. [For the related existence theorem, see Magill (1981)]. The uniqueness implies that the optimal path is monotone (Hartle 1987). Suppose $f'(0) < \rho$. Then, there is no interior OSS. The optimal path converges to the zero stock level ($x = 0$) or to the carrying capacity ($x = K$). However, $x = K$ is not an

OSS. Therefore, we conclude that every optimal path converges to $x = 0$. Notice that a strictly decreasing optimal path implies $c^*(t) > f(x^*(t)) > 0$ and thus the optimal control $c^*(t)$ is interior as far as $x^*(t) > 0$. Therefore, the maximum condition $\lambda(t) = u'(c^*(t))$ (Eq. 3a in the main text) holds over the periods such that $x^*(t) > 0$. Let $T \in [0,\infty]$ be the first time $x^*(t)$ reaches the zero stock level. Then, from the adjoint Eq. (3b),

$$\lim_{t \to T} \dot{\lambda}^*(t) = \lim_{t \to T} \lambda^*(t)[\rho - f'(x^*(t))] > \lambda^*(0)(\rho - f'(0)) > 0, \tag{A1}$$

where the strict inequality follows from $\rho > f'(0) > f'(x)$ for $x > 0$ and $\lim_{t \to T} \lambda(t) > \lambda(0)$ (since $\dot{\lambda}(t) > 0$). Suppose $T = \infty$. Then, (A1) implies $\lim_{t \to T} \lambda^*(t) = \infty$. However, this contradicts the maximum condition (3a), $u'(c^*(t)) = \lambda^*(t)$, since the marginal utility is finite ($\lim_{c \to 0} u'(c) < \infty$) by Assumption 3. Therefore, $T < \infty$, i.e., the resource is optimally exhausted in a finite time.

References

Akao K (2001) Some results for resource games. Institute for Research in Contemporary Political and Economic Affairs Working Paper 2009, Waseda University, Tokyo

Akao K (2004) Tax schemes in a class of differential games. School of Social Sciences Working Paper, Waseda University, Tokyo

Akao K, Farzin YH (2006) When is it optimal to exhaust a resource in a finite time? FEEM Working Paper 23.06, Fondazione Eni Enrico Mattei, Milano

Beddington JR, Watts CMK, Wright WDC (1975) Optimal cropping of self-reproducible natural resources. Econometrica 43:789–802

Berck P (1979) Open access and extinction. Econometrica 47:877–882

Carlowitz HC von (1713) Sylvicultura oeconomica. Brauns, Leipzig

Clark CW (1976) Mathematical bioeconomics: the optimal management of renewable resources. Wiley, New York

Dasgupta P, Heal G (1974) The optimal depletion of exhaustible resources. Review of economic studies. Symposium on the economics of exhaustible resources, pp 3–28

Dasgupta P, Mäler KG (1997) The resource-basis of production and consumption: an economic analysis. In: Dasgupta P, Mäler KG (eds) The environment and emerging development issues, vol1. Clarendon Press, Oxford, pp 1–32

Dechert D, Nishimura K (1983) A complete characterization of optimal growth paths in an aggregate model with a nonconvex production function. J Econ Theory 31:332–354

Deissenberg C, Feichtinger G, Semmler W, Wirl F (2001) History dependence and global dynamics in models with multiple equilibria. Center for Empirical Macroeconomics Working Paper 12, University of Bielefeld, Bielefeld

Farzin YH (2004) Is an exhaustible resource economy sustainable? Rev Dev Econ 8:33–46

Farzin YH, Akao K (2004) Non-pecuniary value of employment and individual labor supply. FEEM Working Paper 158.04, Fondazione Eni Enrico Mattei, Milano

Farzin YH, Akao K (2006) Non-pecuniary value of employment and natural resource extinction. FEEM Working Paper 24.06, Fondazione Eni Enrico Mattei, Milano

Hardin G (1968) The tragedy of commons. Science 162:1243–1247

Hartl RF (1987) A simple proof of the monotonicity of the state trajectories in autonomous control problems. J Econ Theory 41:211–215

Hartl RF, Kort PM, Feichtinger G, Wirl F (2004) Multiple equilibria and threshold due to relative investment costs. J Optim Theory Appl 123:49–82

Heal G (1998) Valuing the future: economic theory and sustainability. Columbia University Press, New York

Levhari D, Liviatan N (1972) On stability in the saddle-point sense. J Econ Theory 4:88–93

Magill MJP (1981) Infinite horizon programs. Econometrica 49:679–712

Nordhaus WD (1994) Reflecting on the concept of sustainable economic growth. In: Pasinetti, LL, Solow RM (eds) Economic growth and the structure of long-term development. Macillan/St. Martin's Press, New York, pp 309–325

Pearce D, Barbier EB, Markandya A. (1990) Sustainable development: economics and environment in the third World. Edward Elgar, Aldershot

Ramsey FP (1928) A mathematical theory of saving. Econ J 38:543–559

Skiba AK (1978) Optimal growth with a convex–concave production function. Econometrica 46:527–539

Solow RM (1998) An almost practical step toward sustainability. Resources for the Future, Washington, D.C.

Sorger G (1998) Markov-perfect Nash equilibria in a class of resources games. Econ Theory 11:78–100

Spence M, Starrett D (1975) Most rapid approach paths in accumulation problems. Int Econ Rev 16:388–403

Tsur Y, Zemel A (2006) Welfare measurement under threats of environmental catastrophes. J Environ Econ Manage 52:421–429

Turner RK, Pearce D, Bateman I (1994) Environmental economics: an elementary introduction. Harvester Wheatsheaf, New York

Vanclay JK (1996) Estimating sustainable timber production from tropical forests. CIFOR Working Paper 11, Center for International Forestry Research, Bogor

World Commission on Environment and Development (1987) Our common future. Oxford University Press, Oxford

Ecol Res (2007) 22: 431–438
DOI 10.1007/s11284-007-0361-4

SPECIAL FEATURE

Yacov Tsur · Amos Zemel

Bio-economic resource management under threats of environmental catastrophes

Received: 6 January 2006 / Accepted: 10 August 2006 / Published online: 28 March 2007
© The Ecological Society of Japan 2007

Abstract We combine ecological and economic dynamics to study the management of a natural resource that supports both ecosystem and human needs. Shrinking the resource base introduces a threat of occurrence of catastrophic ecological events, such as sudden ecosystem collapse. The occurrence conditions involve uncertainty of various types, and the distinction among these types is important for optimal resource management. When uncertainty is due to our ignorance of some aspects of the underlying ecology, the isolated equilibrium states characterizing optimal exploitation for many renewable resource problems become equilibrium intervals. Genuinely stochastic events shift the optimal equilibrium states, but maintain the structure of isolated equilibria.

Keywords Ecosystem dynamics · Resource management · Event uncertainty · Biodiversity · Extinction

Introduction

Recent chronicles are marked by a series of freak environmental events of catastrophic dimensions. Tsunami waves, hurricanes, floods, earthquakes, and extended droughts have stricken various parts of the globe with devastating intensity, inflicting tremendous loss of human lives and impairing the livelihood of millions of people. Threats of deadly epidemic eruptions are also the source of universal concern. The reports on these events have naturally focused on their humanitarian aspects, but it is clear that some of the events also bear significant long-term ecological consequences, including habitat destruction and biodiversity loss.

In some cases the events are the outcome of natural processes that are not affected by human activities, and their policy implications concern mainly the steps required to mitigate the damage. Often, however, anthropogenic pressures on natural resources enhance the threat of occurrence of detrimental events. A case in mind is climate change exacerbated by greenhouse gases released by intensive use of fossil fuels. It is believed that this process can act as a major cause of extinction of numerous animal and plant species (Peterson et al. 2002; Thomas et al. 2004 and references therein). Other important examples are discussed below.

It is clear that a responsible management of our natural resources must account for these risks, which should be properly weighted against the benefits derived from resource exploitation. This work adopts an economic perspective in which human welfare is the dominant consideration. Ecosystem services, then, are valued according to their contribution to human well-being (Heal 2000; Limburg et al. 2002; Brock and Xepapadeas 2003). Often, the affected species do not contribute directly to economic production, but their diminution or extinction entails a loss due to use and nonuse values as well as the loss of option for future benefits such as the development of new medicines (Littell 1992; Bird 1991) or crop resistance (Chichilnisky and Heal 1998). Economic valuations of biodiversity also emphasize its insurance role, building on the ecological premise that genetically rich ecosystems are more resilient and less prone to productivity loss or collapse as the environmental conditions change (Brock and Xepapadeas 2003).

We study the management of a natural resource that serves a dual purpose. First, it provides inputs for

Y. Tsur
Department of Agricultural Economics and Management,
The Hebrew University of Jerusalem, P.O. Box 12,
Rehovot 76100, Israel
E-mail: tsur@agri.huji.ac.il

A. Zemel (✉)
Department of Solar Energy and Environmental Physics,
The Jacob Blaustein Institutes for Desert Research,
Ben Gurion University of the Negev,
Sede Boker Campus 84990, Israel
E-mail: amos@bgu.ac.il

A. Zemel
Department of Industrial Engineering and Management,
Ben Gurion University of the Negev, Beer Sheva 84105, Israel

human production activities and is therefore being exploited for beneficial use, however defined. Second, it supports the existence of other species. Large-scale exploitation competes with the needs of the wildlife populations and, unless controlled, can severely degrade the ecological conditions and lead to extinction and biodiversity loss. Examples for such conflicts abound, including: (1) water diversions for irrigation, industrial or domestic use reduce in-stream flows that support the existence of various fish populations; (2) reclamation of swamps and wetlands that serve as habitat for local plant, bird and animal populations and as "rest areas" for migrating birds (Czech and Parsons 2002); (3) large-scale deforestation (Achard et al. 2002) reduces the living territory of a large number of species, exposing them to risk of extinction (Brooks et al. 1997, 1999); (4) intensive pest control by farmers may entail a take-over by an immune pest species that is harder to control (Hueth and Regev 1974); (5) overgrazing interferes in grass-tree competition in the savannas, pushing towards states of lower productivity (Walker et al. 1981). Overgrazing is also claimed to induce soil erosion and fertility loss over vast semi-arid areas, accelerating desertification processes (Tsoar 1990; Villamil et al. 2001); (6) airborne industrial pollution falls as acid rain on lakes and rivers and interferes with freshwater ecosystems (Jeffries et al. 2003); (7) phosphorus loading into lakes due to agricultural use of fertilizers along the shores can induce an irreversible transition from the oligotrophic (clear) state into a eutrophic (turbid) state (Harper 1992; Carpenter et al. 1999), which severely degrades the value of the lake for fishing and recreation. A similar process is believed to endanger the future of the Baltic Sea (Jansson and Velner 1995).

Biodiversity loss is a good example of the issue under consideration. This process is often induced by a sudden collapse of the ecosystem that shifts the underlying ecology from the current species-rich regime to a new, species-poor regime. This is so because ecosystems are inherently complex, and their nonlinear dynamics can give rise to instabilities, sensitivity to various thresholds and hysteresis phenomena (Holling 1973; Ludwig et al. 1978, 1997; see also Mäler 2000; Limburg et al. 2002; Brock and Starrett 2003; Dasgupta and Mäler 2003 for an economic perspective). We refer to the occurrence of a sudden system collapse as an ecological event.

When the degradation process is gradual and can be monitored and controlled by adjusting exploitation rates, and/or when it involves a discrete ecological event whose occurrence conditions are a-priori known, it is possible to avoid the damage by ensuring that the event will never occur. Often, however, the time of occurrence cannot be predicted in advance because the conditions that trigger ecological events involve uncertainty of various types. The present study characterizes optimal resource exploitation policies under the threat of such events.

Impacts of event uncertainty on resource exploitation policies have been studied in a variety of situations and

contexts, including emission-induced events (Cropper 1976; Clarke and Reed 1994; Tsur and Zemel 1996, 1998b; Aronsson et al. 1998), forest fires (Reed 1984; Yin and Newman 1996), species extinction (Reed 1989; Tsur and Zemel 1994), seawater intrusion into coastal aquifers (Tsur and Zemel 1995, 2004), and political crises (Long 1975; Tsur and Zemel 1998a). The hovering risk typically leads to prudence and conservation, but may also invoke the opposite effect, encouraging aggressive exploitation in order to derive maximal benefit prior to occurrence (Clarke and Reed 1994).

Tsur and Zemel (1998b, 2004, 2007) trace these apparently conflicting results to different assumptions concerning the event occurrence conditions and the ensuing damage they inflict. An important distinction relates to the source of uncertainty: stochastic environmental conditions or partial ignorance (on the part of the planner) of key system parameters. We show that this distinction bears important implications for optimal exploitation policies and alters properties that are considered standard. For example, the optimal stock processes of renewable resources typically approach isolated equilibrium states. This feature, it turns out, no longer holds under ignorance-related uncertainty: the equilibrium point expands into an equilibrium interval whose size depends on the expected loss, and the eventual steady state is determined by the initial stock. In contrast, stochastic events maintain the structure of isolated equilibria and the effect of uncertainty is manifest via the shift it induces on the equilibrium states.

In this paper, we avoid detailed exposition and mathematical derivations of optimal resource management under uncertainty [these are presented in Tsur and Zemel (2004, 2007) and the references they cite]. Our aim here is to explain the economic reasoning and show how ecological dynamics (manifest via abrupt event occurrence at some unknown future time) combine with economic considerations to characterize optimal exploitation policies under threats of environmental events.

The management problem

Resource economics analysis typically revolves around tradeoffs and balances. To determine optimal exploitation policies one needs to weigh the benefits derived from the use of the resource against the associated costs. The tradeoffs take various forms depending on the specific context. In the simpler cases, one compares the diminishing marginal benefits from resource use to the increasing cost or damage implied by this use. The solution to this static optimization problem determines optimal extraction, beyond which further exploitation is not worthwhile, giving rise to 'economic depletion'. In other scenarios, the tradeoff is between current benefits and future scarcity. Problems of this type are analyzed using dynamic optimization techniques and attention is focused on the physical depletion of the resource: under what conditions is depletion desirable, and when should depletion take place? The considerations of the present work are intertemporal, but the tradeoff is between current benefits and the increasing hazard of a hovering environmental catastrophe.

We consider the management of some environmental resource that is essential to maintain a functioning ecosystem and at the same time is exploited for human production activities. The stock S of the resource can represent the uncultivated area of land of potential agricultural use, the water level at some lake or river or the degree of cleanliness (measured, e.g., by atmospheric concentrations or as the pH level of a lake affected by acid rain or polluting effluents). Without human interference, the stock dynamics are determined by the natural regeneration rate $G(S)$ (corresponding to groundwater recharge, to the decay rate of a pollution stock or to the natural expansion rate of a forest area). The functional form of G depends on the particular resource under consideration, but we assume the existence of some upper bound $\bar{S}$ for the stock, corresponding to the resource carrying capacity, such that $G(\bar{S}) = 0$ and $G'(\bar{S}) \leqslant 0$. With x_t representing the rate of resource exploitation, the resource stock evolves with time according to

$$\mathrm{d}S_t/\mathrm{d}t = G(S_t) - x_t. \tag{2.1}$$

Exploitation at a rate x entails several consequences. First, it generates a benefit flow at the rate $Y(x)$ (from the use of land, water or timber or from the economic activities that involve the emission of pollutants). Second, it bears the exploitation cost $C(S)x$, where the unit cost $C(S)$ can depend on the resource stock. Third, reducing the stock level [by setting $x > G(S)$] entails decreasing the value of the services derived from the ecosystem that depends on the same resource for its livelihood. This loss of value is expressed in terms of the damage rate $D(S)$. The net benefit flow is then given by $Y(x) - C(S)x - D(S)$.

Moreover, a decrease in the resource stock S increases the probability of occurrence of an influential event of adverse consequences due to the abrupt collapse of the ecosystem it supports. In some cases the event is triggered when S crosses an a priori unknown critical level, which is revealed only when occurrence actually takes place. Alternatively, the event may be triggered at any time by stochastic external effects (such as unfavorable weather conditions or the outburst of some disease). Since the resilience of the ecosystem depends on the current resource stock, the occurrence probability also depends on this state. We refer to the former type of uncertainty—that due to our ignorance regarding the conditions that trigger the event—as endogenous uncertainty (signifying that the event occurrence is solely due to the exploitation decisions) and to the latter as exogenous uncertainty. Both types of uncertainty imply that the occurrence

time cannot be predicted in advance. Nevertheless, it turns out that the optimal policies are sensitive to the distinction between these types.

Let T denote the (random) event occurrence time, such that $[0, T]$ and (T, ∞) are the *pre-event* and *post-event* periods, respectively. The benefit $Y(x_t) - C(S_t)x_t - D(S_t)$ defined above is the pre-event net benefit flow at time $t < T$. Let $\varphi(S_T)$ denote the post-event value at the occurrence time T, consisting of the present value generated by the optimal post-event policy from time T onward (discounted to time T) as well as the immediate consequences of the event (see examples below).

An exploitation policy $\{x_t, t \geq 0\}$ gives rise to the resource process $\{S_t, t \geq 0\}$ via (2.1) and generates the expected present value

$$E_T \left\{ \int_0^T [Y(x_t) - C(S_t)x_t - D(S_t)]e^{-rt}\mathrm{d}t + e^{-rT}\varphi(S_T) | T > 0 \right\} \tag{2.2}$$

where E_T denotes the expectation operator with respect to the probability distribution of T and r is the time rate of discount (for extended discussions on the choice of the discount rate, see the collection of works edited by Portney and Weyant 1999). The distribution of T and the ensuing conditional expectation depend on the type of uncertainty and on the exploitation policy. Given the initial stock S_0, we seek the feasible policy that maximizes (2.2) subject to (2.1). In the next section, we consider the reference case in which T can be predicted prior to occurrence and characterize the optimal policy for this case. Endogenous and exogenous uncertainty are discussed in Sects. 4 and 5, respectively.

Predictable occurrence time

Suppose that driving the stock to some *known* critical level S_c triggers ecosystem collapse, which entails an immediate damage (penalty) $\psi > 0$ and prohibits any further decrease of the resource stock. Given that the critical state S_c has been reached, the optimal post-event policy is to maintain the stock at that level and the corresponding post-event value is $\varphi(S_c) = W(S_c) - \psi$, where

$$W(S) = [Y(G(S)) - C(S)G(S) - D(S)]/r \tag{3.1}$$

is the present value generated by the steady-state policy that sets the exploitation rate at the natural regeneration rate $G(S)$. The post-event value $\varphi(S_c)$, thus, accounts both for the fact that the stock cannot be further decreased (to avoid further damage) and for the penalty implied by occurrence. The event is triggered at the critical level S_c, hence the occurrence time T is defined by the condition $S_T = S_c$ ($T = \infty$ if the stock is always kept above S_c).

Since T is subject to choice, the conditional expectation in (2.2) can be ignored and the management problem becomes

$$V^c(S_0) = \text{Max}_{\{T, x_t\}} \left\{ \int_0^T [Y(x_t) - C(S_t)x_t \right.$$

$$\left. - D(S_t)]e^{-rt}dt + e^{-rT}\varphi(S_T) \right\} \quad (3.2)$$

subject to (2.1), $x_t \geq 0$; $S_T = S_c$ and $S_0 > S_c$ given. Optimal processes associated with this "certainty" problem are indicated with a "c" superscript. Occurrence is evidently undesirable, since just above S_c it is preferable to extract at the regeneration rate and enjoy the benefit flow $rW(S_c)$ associated with it rather than trigger the event and bear the penalty ψ. Thus, the event should be avoided, the stock is kept above the critical level for all t and $T = \infty$. The certainty problem, thus, is reformulated as

$$V^c(S_0) = \text{Max}_{\{x_t\}} \int_0^\infty [Y(x_t) - C(S_t)x_t - D(S_t)]e^{-rt}dt \quad (3.3)$$

subject to (2.1), $x_t \geq 0$; $S_t > S_c$ and S_0 given. The effect of the known critical stock enters only via the lower bound imposed on the stock process. This simple problem is akin to standard resource management problems and can be treated by a variety of optimization methods (see, e.g., Tsur and Zemel 1994, 1995, 2004). Here, we review the main properties of the optimal plan.

We note first that because problem (3.3) is autonomous (time enters explicitly only through the discount factor) the optimal stock process S_t^c evolves monotonically in time. The property is based on the observation that if the process reaches the same state at two distinct times, then the planner faces the same optimization problem at both times. This rules out the possibility of the optimal stock process exhibiting a local maximum, because the conflicting decisions to increase the stock (before the maximum) and decrease it (after the maximum) are taken at the same stock levels. Similar considerations exclude a local minimum. Since S_t^c is monotone and bounded in $[S_c, \bar{S}]$, it must approach a steady state in this interval. Using the variational method of Tsur and Zemel (2001), possible steady states are located by means of a simple function $L(S)$ of the state variable, which is determined by the model specifications. In particular, an internal state S in $(S_c, \bar{S})$ can qualify as an optimal steady state only if it is a root of $L(\cdot)$, i.e. $L(S) = 0$, while the corners S_c or $\bar{S}$ can be optimal steady states only if $L(S_c) \leq 0$ or $L(\bar{S}) \geq 0$, respectively.

For the problem at hand, we find that when $Y'(0) < C(\bar{S})$, exploitation is never profitable. In this case $L(\bar{S}) > 0$ and the unexploited stock eventually settles at the maximum level $\bar{S}$. The condition for convergence to

the other corner solution (the critical level S_c) is discussed below.

Under the appropriate curvature assumptions, $L(S)$ has a unique root $\hat{S}^c$ in $[S_c, \bar{S}]$. In this case, $\hat{S}^c$ is the unique steady state to which the optimal state process S_t^c converges monotonically from any initial stock.

When the function $L(S)$ obtains a root above the critical state S_c, the constraint $S_t > S_c$ is never binding because there is no advantage in shrinking the stock below the steady state. Thus, the risk of occurrence has no effect on the optimal policy. However, with $S_c > \hat{S}^c$ a process approaching the root of $L(S)$ must cross the critical state and trigger the event, which cannot be optimal. The optimal stock process S_t^c, then, converges *monotonically and asymptotically* to a steady state at S_c. By keeping the process above the no-event optimal (i.e., the optimal policy without the constraint $S_t > S_c$), the event threat imposes prudence and a lower rate of extraction.

While the discussion above implies that the stock process must approach a steady state, the time to enter this state is a choice variable. Using the conditions for an optimal entry time, one finds that the optimal extraction rate x_t^c smoothly approaches the steady state regeneration rate and the approach of S_t^c towards the steady state is asymptotic, i.e., the optimal stock process will not enter the steady state at a finite time. These properties, as well as the procedure to obtain the full-time trajectory of the optimal plan, are derived in Tsur and Zemel (2004).

The event in this formulation is never triggered, and the exact value of the penalty is irrelevant (so long as it is positive). This result is due to the requirement that the post-event stock is not allowed to decrease below the critical level. In fact, this requirement can be relaxed whenever the penalty is sufficiently large to deter triggering the event in any case. The lack of sensitivity of the optimal policy to the details of the catastrophic event is evidently due to the ability to avoid the event occurrence altogether. This may not be feasible (or optimal) when the critical stock level is not a-priori known. The optimal policy may, in this case, lead to unintentional occurrence, whose exact consequences must be accounted for in advance. We turn, in the following two sections, to analyzing the effect of uncertain catastrophic events on resource management policies.

Endogenous events

A catastrophic event is called endogenous if its occurrence is determined solely by the resource exploitation policy, although the exact threshold level S_c at which the event is triggered is not a-priori known and the event occurrence time, for a given exploitation policy, cannot be predicted in advance. This type of uncertainty, however, allows avoiding the occurrence risk altogether by keeping the resource stock at or above its initial state

S_0. The post-event value is specified again as $\varphi(S) = W(S) - \psi$.

Let $F(S) = Pr\{S_c \leq S\}$ and $f(S) = dF/dS$ denote the probability distribution and density functions of the critical level S_c and denote by $q(S)$ the conditional density of occurrence due to a small stock decrease given that the event has not occurred by the time the state S was reached

$$q(S) = f(S)/F(S). \qquad (4.1)$$

We assume that $q(S)$ does not vanish in the relevant range, hence no state below the initial stock can be considered a-priori safe.

The distribution of the threshold S_c induces a distribution on the occurrence time T in a nontrivial way, which depends on the exploitation history. To see this notice that as the stock process evolves in time, the distributions of S_c and T are modified since at time t it is known that S_c must lie below the lowest state so far, $\tilde{S}_t = \text{Min}_{0 \leq \tau \leq t}\{S_\tau\}$ (otherwise the event would have occurred at some time prior to t). Thus, the distributions of S_c and T involve the entire history up to time t, which complicates the evaluation of the conditional expectation in (2.2). It appears, therefore, that (2.2) is not a proper formulation of a dynamic optimization problem. However, the situation is simplified when the stock process S_t evolves monotonically in time, since then $\tilde{S}_t = S_0$ if the process is non-decreasing (and no information relevant to the distribution of S_c is revealed) and $\tilde{S}_t = S_t$ if the process is non-increasing (and all the relevant information is given by the current stock S_t).

As in the case of a known threshold, the *optimal* stock process evolves monotonically in time also under uncertainty. This property extends the reasoning of the certainty case above: if the process reaches the same state at two different times, and no new information on the critical level has been revealed during that period, then the planner faces the same optimization problem at both times. This rules out the possibility of a local maximum for the optimal state process, because $\tilde{S}_t$ remains constant around the maximum, yet the conflicting decisions to increase the stock (before the maximum) and decrease it (after the maximum) are taken at the same stock levels. A local minimum can also be ruled out even though the decreasing process modifies $\tilde{S}_t$ and adds information on S_c. However, it cannot be optimal to decrease the stock under occurrence risk (prior to reaching the minimum) and then increase it with no occurrence risk (after the minimum) from the same state. For a complete proof, see Tsur and Zemel (1994).

For a non-decreasing stock process it is known in advance that the event will never occur and the uncertainty problem reduces to the certainty problem (3.3) corresponding to $S_c = 0$ (the latter can be referred to as the 'non-event' problem because the event cannot be triggered; see Tsur and Zemel 2004). For non-increasing stock process the distribution of T is obtained from the distribution of S_c as follows:

$$1 - F_T(t) \equiv Pr\{T > t | T > 0\} = Pr\{S_c < S_t | S_c < S_0\}$$
$$= F(S_t)/F(S_0). \qquad (4.2)$$

Using this T-distribution, the conditional expectation (2.2) can be evaluated for non-increasing state processes, yielding the following management problem

$$V^{\text{aux}}(S_0) = \max_{\{x_t\}}\left\{\int_0^\infty \{Y(x_t) - C(S_t)x_t - D(S_t)\right.$$
$$\left. + q(S_t)[x_t - G(S_t)]\varphi(S_t)\}\frac{F(S_t)}{F(S_0)}e^{-rt}dt\right\} \qquad (4.3)$$

subject to (2.1), $x_t \geq 0, S_t \geqslant \hat{S}^c$ and S_0 given. This problem is referred to as the *auxiliary* problem, and the associated optimal processes are denoted by the superscript *aux*.

Formulated as an autonomous problem, the auxiliary problem also gives rise to an optimal stock process that converges monotonically to a steady state. We find that the associated steady state $\hat{S}^{\text{aux}}$ represents a higher resource stock than the steady state $\hat{S}^c$ corresponding to the certainty problem, and the difference depends on the quantity $q(S)r\psi$ that measures the expected loss due to an infinitesimal decrease in stock. (The event inflicts an instantaneous penalty ψ, or equivalently, a permanent loss flow at the rate $r\psi$, that could have been avoided by the safe policy of keeping the stock at the current level S.)

Notice that at this stage it is not clear whether the uncertainty problem at hand reduces to the certainty problem or to the auxiliary problem, since it is not a priori known whether the optimal stock process decreases with time. In order to determine the optimal process S_t^{en} implied by endogenous events, we compare the trajectories of the auxiliary problem with those obtained with the certainty problem corresponding to $S_c = 0$. The following characterization holds:

(a) S_t^{en} increases at stock levels below $\hat{S}^c$ (coinciding with the certainty process S_t^c).
(b) S_t^{en} decreases at stock levels above $\hat{S}^{\text{aux}}$ (coinciding with the auxiliary process S_t^{aux}).
(c) All stock levels in $[\hat{S}^c, \hat{S}^{\text{aux}}]$ are equilibrium states of S_t^{en}.

The equilibrium interval is unique to optimal stock processes under endogenous uncertainty. Its boundary points attract any process initiated outside the interval while processes initiated within it must remain constant. This feature is evidently related to the splitting of the intertemporal exploitation problem to two distinct optimization problems depending on the initial trend of the optimal stock process. At $\hat{S}^{\text{aux}}$, the expected loss due to occurrence is so large that entering the interval cannot be optimal even if under certainty extracting above the regeneration rate would yield a higher benefit. Within the equilibrium interval it is possible to eliminate the occurrence risk altogether by not reducing the stock below its current level. As we shall see below, this possibility is not available for exogenous events that do not give rise to equilibrium intervals.

Endogenous uncertainty implies more conservative exploitation as compared with the certainty case. Observe that the steady state $\hat{S}^{aux}$ is a planned equilibrium level. In actual realizations, the process may be interrupted by the event at a higher stock level, and the actual equilibrium level in such cases will be the realized critical state S_c.

A feature similar to both the certain event and the endogenous event cases is the smooth transition to the steady states. When the initial stock is outside the equilibrium interval, the condition for an optimal entry time to the steady state implies that extraction converges smoothly to the recharge rate and the planned steady state will not be entered at a finite time. It follows that when the critical level actually lies below $\hat{S}^{aux}$,, uncertainty will never be resolved and the planner will never find out if the adopted policy of approaching $\hat{S}^{aux}$ is indeed safe. Of course, in the less fortunate case in which the critical level lies above the steady state, the event will occur at finite time and the damage will be inflicted.

Exogenous events

Ecological events that are triggered by environmental conditions beyond the planners' control are termed 'exogenous'. Changing the resource stock level can modify the hazard of immediate occurrence through the effect of the stock on the resilience of the ecosystem, but no exploitation policy is completely safe since the collapse event is triggered by stochastic changes in exogenous conditions. This type of event uncertainty has been applied for the modeling of a variety of resource-related situations, including nuclear waste control (Cropper 1976; Aronsson et al. 1998), environmental pollution (Clarke and Reed 1994; Tsur and Zemel 1998b) and groundwater resource management (Tsur and Zemel 2004). Here, we consider the implications for biodiversity conservation. Under exogenous event uncertainty, the fact that a certain stock level has been reached in the past without triggering the event does not rule out occurrence at the same stock level sometime in the future, when exogenous circumstances turn out to be less favorable. Therefore, the mechanism that gives rise to equilibrium intervals under endogenous uncertainty does not work here.

As above, the post-event value is denoted by $\varphi(S)$ and the expected present value of an exploitation policy that can be interrupted by an event at time T is given in (2.2). The probability distribution of T, $F(t) = \Pr\{T \leq t\}$, is defined in terms of a stock-dependent hazard rate function $h(S)$ satisfying

$$h(S_t) = f(t)/[1 - F(t)] = -\mathrm{d}\{\log[1 - F(t)]\}/\mathrm{d}t, \quad (5.1)$$

hence

$$F(t) = 1 - \exp[-\Omega(t)] \quad \text{and} \quad f(t) = h(S_t)\exp[-\Omega(t)],$$
$$(5.2)$$

where

$$\Omega(t) = \int_0^t h(S_\tau)\mathrm{d}\tau \quad (5.3)$$

can be considered as a cumulative 'hazard stock'. With a state-dependent hazard rate, the quantity $h(S_t)\mathrm{d}t$ measures the conditional probability that the event will occur during the infinitesimal interval $(t, t + \mathrm{d}t)$ given that it has not occurred by time t when the stock level is S_t.

We assume that no stock level is completely safe, hence $h(S)$ does not vanish and $\Omega(t)$ diverges for any feasible stock process as $t \to \infty$. We further assume that $h(S)$ is decreasing, because a shrinking stock deteriorates ecosystem conditions and increases the hazard for environmental collapse.

Given the distribution of T, the management problem (2.2) is formulated as

$$V^{ex}(S_0) = \max_{\{x_t\}} \int_0^\infty [Y(x_t) - C(S_t)x_t$$
$$- D(S_t) + h(S_t)\varphi(S_t)]e^{-rt - \Omega(t)}\mathrm{d}t \quad (5.4)$$

subject to (2.1), $x_t \geq 0$; $S_t \geq 0$ and S_0 given. Unlike the auxiliary problem (4.3) used above to characterize decreasing policies under endogenous events, problem (5.4) provides the correct formulation for exogenous events regardless of whether the stock process decreases or increases. We use the superscript 'ex' to denote variables associated with the exogenous uncertainty problem (5.4).

The explicit time dependence of the distribution $F(t)$ of (5.2) renders formulation (5.4) of the optimization problem non-autonomous. [Note the presence of the hazard stock $\Omega(t)$ in the effective discount factor]. Nevertheless, the argument for the monotonic behavior of the optimal stock process S_t^{ex} holds, and the associated steady states can be derived (see Tsur and Zemel 1998b).

When the event corresponds to species extinction, it can occur only once and the loss is irreversible. If a further reduction in the hazard-mitigating stock is forbidden, the steady state $\hat{S}^{ex}$ must lie above the certainty equilibrium $\hat{S}^c$, implying more prudence and conservation compared to the policy free of uncertainty.

Biodiversity conservation considerations enter via the shift in steady states, which measures the marginal expected loss due to a small decrease in the resource stock. The latter implies a higher occurrence risk, which in turn calls for a more prudent exploitation policy. Indeed, if the hazard is stock-independent ($h'(S) = 0$), the resulting steady states coincide. In this case, exploitation has no effect on the expected loss, hence the tradeoffs that determine the optimal equilibrium need not account for the hazard, regardless of how severe the damage may be. For a decreasing hazard function ($h'(S) < 0$), however, the degree of prudence (as measured by the difference $\hat{S}^{ex} - \hat{S}^c$) increases with the penalty ψ.

The requirement that the stock must not be further reduced following occurrence can be relaxed. For this situation, the post-event value is specified as $\varphi(S) = V_c(S) - \psi$, yielding a more complex expression for the steady states, but the property $\hat{S}^{ex} > \hat{S}^c$ remains valid (Tsur and Zemel 1998b).

Another interesting situation involving exogenous events arises when the damaged ecology can be restored at the cost ψ. For example, the extinct population may not be endemic to the inflicted region and can be renewed by importing individuals from unaffected habitats. When restoration is possible, event occurrence inflicts a penalty, but does not affect the hazard of future events. For this case, we find that the shift in equilibrium states depends on $d[\psi h(S)]/dS$, which measures the sensitivity of the expected damage to small changes in stock.

When the event penalty ψ also depends on the stock S, policy implications become more involved. Curiously, the case of increasing $\psi(S)$ and constant hazard implies more vigorous exploitation, compared to the risk-free environment (Clarke and Reed 1994). An example for this situation is the case of a "doomsday" event (following which the ecosystem is ruined forever, so that no post-event benefit can be derived) induced by an earthquake or a volcanic eruption (hence the corresponding hazard is independent of S). In this case, the damage equals the overall value of the ecosystem, which increases with the resource stock. As the rate of extraction does not affect the occurrence probability, the hovering threat encourages enhanced extraction in order to enjoy as much benefit as is possible prior to occurrence.

Concluding comments

Exploitation of natural resources is typically considered in the context of their direct contribution to human activities, while their roles in supporting ecological needs are often overlooked in the economic analysis. In this work, we examine ways to incorporate ecological considerations within resource exploitation models. We focus on threats of abrupt ecological events whose occurrence inflicts a penalty due to an adverse change in the ecosystem regime. Unlike gradual changes (time-varying costs and damage, stochastic regeneration processes, etc.), which allow adaptation and updating the exploitation policy in response to the changing conditions, abrupt event uncertainty is resolved only upon occurrence, when policy changes cannot prevent the damage. Thus, the expected loss must be fully accounted for prior to occurrence, with significant modifications to the optimal resource management rules.

We distinguish between two types of events that differ in the conditions that trigger their occurrence. An endogenous event occurs when the resource stock crosses an uncertain threshold level, while exogenous events are triggered by coincidental random environmental conditions. We find that the optimal exploitation policies are sensitive to the type of the threatening events. Under endogenous events, the optimal stock process approaches the nearest edge of an equilibrium interval or remains fixed if the initial stock lies inside the equilibrium interval. The eventual equilibrium stock depends on the initial conditions. In contrast, the equilibrium states under exogenous uncertain events are singletons that attract the optimal processes from any initial stock. The shift of these equilibrium states relative to their certainty counterparts is due to the marginal expected loss associated with the events and serves as a measure of how much prudence they imply. In most cases, the hovering threat encourages conservation.

A feature common to all the events considered here is that information accumulated in the course of the process regarding occurrence conditions does not affect the original policy until the time of occurrence. In some situations, however, it is possible to learn during the process and continuously update estimates of the occurrence probability. This possibility introduces another consideration to the tradeoffs that determine optimal exploitation policies. In this case one has to account also for the information content regarding occurrence probability associated with each feasible policy. While learning and expectations have been incorporated within economic models of gradual environmental damage (see Karp and Zhang 2006), the investigation of these more complicated models in the context of abrupt events is yet to be undertaken.

References

Achard F, Eva HD, Stibig H-J, Mayaux Ph, Gallego J, Richards T, Malingreau J-P (2002) Determination of deforestation rates of the world's humid tropical forests. Science 297:999–1002

Aronsson T, Backlund K, Löfgren KG (1998) Nuclear power, externalities and non-standard Piguvian taxes: a dynamical analysis under uncertainty. Environ Resour Econ 11:177–195

Bird C (1991) Medicines from the forest. New Sci 17:34–39

Brock WA, Starrett D (2003) Managing systems with non-convex positive feedback. Environ Resour Econ 26:575–602

Brock WA, Xepapadeas A (2003) Valuing biodiversity from an economic perspective: a unified economic, ecological and genetic approach. Am Econ Rev 93:1597–1614

Brooks TM, Pimm SL, Collar NJ (1997) Deforestation predicts the number of threatened birds in insular Southeast Asia. Conserv Biol 11:382–394

Brooks TM, Pimm SL, Oyugi JO (1999) Time lag between deforestation and bird extinction in tropical forest fragments. Conserv Biol 13:1140–1150

Carpenter S, Ludwig D, Brock W (1999) Management of eutrophication for lakes subject to potentially irreversible change. Ecol Appl 9:751–771

Chichilnisky G, Heal G (1998) Economic returns from the biosphere. Nature 39:629–630

Clarke HR, Reed WJ (1994) Consumption/pollution tradeoffs in an environment vulnerable to pollution-related catastrophic collapse. J Econ Dyn Control 18:991–1010

Cropper ML (1976) Regulating activities with catastrophic environmental effects. J Environ Econ Manage 3:1–15

Czech HA, Parsons KC (2002) Agricultural wetlands and waterbirds: a review. Waterbirds 25:56–65

Dasgupta P, Mäler K-G (2003) The economics of non-convex ecosystems: introduction. Environ Resour Econ 26:499–525

Harper D (1992) Eutrophication of freshwater. Chapman and Hall, London

Heal G (2000) Nature and the marketplace: capturing the value of ecosystem services. Island Press, Washington

Holling CS (1973) Resilience and stability of ecological systems. Annu Rev Ecol Syst 4:1–23

Hueth D, Regev U (1974) Optimal agricultural pest management with increasing pest resistance. Am J Agric Econ 56:543–553

Jansson B-O, Velner H (1995) The Baltic: the sea of surprises. In: Gunderson LH, Holling CS, Light SS (eds) Barriers and bridges to the renewal of ecosystems and institutions. Columbia University Press, New York, pp 292–372

Jeffries DS, Brydges TG, Dillon PJ, Keller W (2003) Monitoring the results of Canada/USA acid rain control programs: some lake responses. Environ Monit Assess 88:3–19

Karp L, Zhang J (2006) Regulation with anticipated learning about environmental damages. J Environ Econ Manage 51:259–279

Limburg KE, O'Neill RV, Costanza R, Farber S (2002) Complex systems and valuation. Ecol Econ 41:409–420

Littell R (1992) Endangered and other protected species: federal law and regulations. The Bureau of National Affairs, Washington

Long NV (1975) Resource extraction under the uncertainty about possible nationalization. J Econ Theor 10:42–53

Ludwig D, Jones D, Holling CS (1978) Qualitative analysis of insect outbreak systems: the spruce budworm and the forest. J Anim Ecol 47:315–332

Ludwig D, Walker B, Holling CS (1997) Sustainability, stability and resilience, Conserv Ecol 1:7 (available online at http://www.consecol.org/vol1/iss1/art7/)

Mäler K-G (2000) Development, ecological resources and their management: a study of complex dynamic systems. Eur Econ Rev 44:645–665

Peterson AT, Ortega-Huerta MA, Bartley J, Sanchez-Cordero V, Soberón J, Buddemeier RH, Stockwell DRB (2002) Future projections for Mexican faunas under global climate change scenarios. Nature 416:626–629

Portney P, Weyant J (eds) (1999) Discounting and intergenerational equity. Resources for the Future, Washington

Reed WJ (1984) The effect of the risk of fire on the optimal rotation of a forest. J Environ Econ Manage 11:180–190

Reed WJ (1989) Optimal investment in the protection of a vulnerable biological resource. Nat Resour Model 3:463–480

Thomas CD, Cameron A, Green RE, Bakkenes M, Beaumont LJ, Collingham YC, Erasmus BFN, de Siqueira MF, Grainger A, Hannah L, Hughes L, Huntley B, van Jaarsveld AS, Midgley GF, Miles L, Ortega-Huerta MA, Peterson AT, Phillips OL, Williams SE (2004) Extinction risk from climate change. Nature 427:145–148

Tsoar H (1990) The ecological background, deterioration and reclamation of desert dune sand. Agric Ecosyst Environ 33:147–170

Tsur Y, Zemel A (1994) Endangered species and natural resource exploitation: extinction vs. coexistence. Nat Resour Model 8:389–413

Tsur Y, Zemel A (1995) Uncertainty and irreversibility in groundwater resource management. J Environ Econ Manage 29:149–161

Tsur Y, Zemel A (1996) Accounting for global warming risks: resource management under event uncertainty. J Econ Dyn Control 20:1289–1305

Tsur Y, Zemel A (1998a) Trans-boundary water projects and political uncertainty. In: Just R, Netanyahu S (eds) Conflict and cooperation on trans-boundary water resources. Kluwer, Dordrecht pp 279–295

Tsur Y, Zemel A (1998b) Pollution control in an uncertain environment. J Econ Dyn Control 22:967–975

Tsur Y, Zemel A (2001) The infinite horizon dynamic optimization problem revisited: a simple method to determine equilibrium states. Eur J Oper Res 131:482–490

Tsur Y, Zemel A (2004) Endangered aquifers: groundwater management under threats of catastrophic events. Water Resour Res 40:W06S20 doi:10.1029/2003WR002168

Tsur Y, Zemel A (2007) Resource exploitation, biodiversity loss and ecological events. In: Kontoleon A, Pascual U, Swanson T (eds) Biodiversity economics, principles, methods and applications. Cambridge University Press, Cambridge (forthcoming)

Villamil MB, Amiotti NM, Plinemann N (2001) Soil degradation related to over grazing in the semiarid Caldenal area of Argentina. Soil Sci 166:441–452

Walker BH, Ludwig D, Holling CS, Peterman RM (1981) Stability of semiarid savanna grazing systems. J Ecol 69:473–498

Yin R, Newman DH (1996) The effect of catastrophic risk on forest investment decisions. J Environ Econ Manage 31:186–197